Telexistence

Telexistence

Susumu Tachi

The University of Tokyo, Japan

&

Keio University, Japan

World Scientific

NEW JERSEY · LONDON · SINGAPORE · BEIJING · SHANGHAI · HONG KONG · TAIPEI · CHENNAI

Published by

World Scientific Publishing Co. Pte. Ltd.

5 Toh Tuck Link, Singapore 596224

USA office: 27 Warren Street, Suite 401-402, Hackensack, NJ 07601

UK office: 57 Shelton Street, Covent Garden, London WC2H 9HE

British Library Cataloguing-in-Publication Data
A catalogue record for this book is available from the British Library.

TELEXISTENCE

ISBN-13 978-981-283-633-5
ISBN-10 981-283-633-0

Typeset by Stallion Press
Email: enquiries@stallionpress.com

Printed in Singapore by Mainland Press Pte Ltd

Preface

Although the term virtual reality (VR) was coined around 1989, similar researches have been carried out since the early 1980s in various fields such as computer interface, computer-aided design and manufacturing, simulation and simulators, interactive art, communication, and robotics.

Since the Engineering Foundation Conference on Human Machine Interfaces for Teleoperators and Virtual Environments was held in Santa Barbara in 1990, such researches have been discussed collectively, and this field of research has been recognized as VR.

The three basic components of VR are life-sized three-dimensional space, real-time interaction, and self-projection.

While the gaming applications of VR were identified early, industrial applications such as car design tools or medical care have been keenly pursued only recently. Following this trend, exhibitions presenting industrial developments, such as the industrial VR show (IVR), have flourished year after year in Japan.

In addition, academic conferences such as the International Virtual Reality Conference hosted by the Computer Society of IEEE, the world's most authoritative academy, are conducted every year, and VR research has occupied the principal focus of the emerging technologies presented at SIGGRAPH, the world's largest conference and exhibition on computer graphics and interaction techniques. Moreover, the Virtual Reality Society of Japan has made advancements every year since it was established in 1996, and VR has been established as a solid research area and discipline in Japan.

In the USA, the potential for advancements in the field of VR is extremely high. On 20 February 2008, for example, the National Science Foundation (NSF) selected VR as one of the "14 grand engineering challenges for the 21st century" that could improve the quality of human life around the world.

In 1980, the author conceptualized and proposed the concept of "telexistence" that enables a highly realistic sensation of existence in another remote place without any actual travel and demonstrated its feasibility through the construction of alter-ego robot systems such as TELESAR, which was developed under the national large-scale project on "Robots in Hazardous Environments," as well as the HRP super-cockpit biped robot system developed under the "Humanoid Robotics Project" of the Ministry of Economy, Trade and Industry (METI).

Thereafter, the author and his team have researched and developed a mutual telexistence system TELESAR II, which can generate the effect of a person in a remote place as directly appearing in local space using a combination of the alter-ego robot and the retro-reflective projection technology (RPT). Furthermore, full-color 360-degree panoramic and autostereoscopic system TWISTER has been developed using a rotating parallax barrier and LED arrays; it does not require any special eyewear such as shutter glasses. Based on the results, the feasibility of mutual telexistence has been demonstrated.

Furthermore, the author has aimed at developing a system that enables recording, communicating, and reconstructing tactual sense. This is required in addition to the visual and auditory senses in order to re-create realistic sensations. The author and his team have pursued this line of thought through the development of a distributed vector-type sensor called GelForce and an electrocutaneous display.

Moreover, augmented reality based on RPT (retro-reflective projection technology), which represents a projection of informational space onto real-world space via the RPT, has been developed for the purpose of improving real-world environments, for example, providing optical camouflage using RPT.

While the majority of the VR researches have focused on the computer-generated world, the author's research is individualized by a fusion of VR and robotics, which is always directed at the real world as exemplified in telexistence.

Telexistence is fundamentally a general technology that enables a human being to have a real-time sensation of being at a place other than where he or she actually exists and being able to interact with the remote environment, which may be real, virtual, or a combination of both. It also refers to an advanced type of teleoperation system that enables an operator at the control to perform remote tasks dexterously with the feeling of existing in a surrogate robot working in a remote environment. Telexistence in the real environment through a virtual environment is also possible. The author believes that telexistence has the potential to release human beings from the restrictions of their cognitive limits and physical constraints.

This book introduces this concept of telexistence, explains how the concept can be realized as a technology, precisely illustrates real examples of the realization of the telexistence technology, and lists prospects and future advancement of telexistence.

A complete review book on the telexistence technology has not been published yet. This is the first book on this emerging technology of telexistence written by the inventor of telexistence.

Susumu Tachi, Ph.D.

Contents

Chapter 1

Virtual Reality and Telexistence

According to the American Heritage Dictionary, "virtual" is defined as existing in effect or in essence though not in actual fact or form. Thus, virtual reality is an entity which contains the essence of reality and is effectively real. It can provide a basis for technology which enables humans to experience events and act in a computer-synthesized environment just as if they were in a real environment. Although telexistence (tel–existence) is essentially the same concept as virtual reality, it takes a different point of view. It represents a new concept, which frees humans from the restrictions of time and space and allows them to be effectively present in places other than their current location as well as to interact with those remote environments, which may be real, computer-synthesized, or a combination of both. Thus, virtual reality and telexistence are essentially the same concept expressed in a different manner. Usually, virtual reality is used for computer-synthesized worlds, while telexistence is used for the real world. However, both concepts can be regarded as tools for communication, control, and creation (the 3Cs) or entertainment, experience/education, and elucidation (the 3Es).

1.1. What is Telexistence and What is Virtual Reality

The concept of telexistence was proposed by the author in 1980 and played the role of the fundamental principle behind the eight-year national large-scale project entitled "Advanced Robot Technology in Hazardous Environments," which was established in 1983 together with the concept of third-generation robotics (Tachi *et al.*, 1980, 1981; Tachi and Abe, 1982; Tachi and Komoriya, 1982).

Telexistence is a concept that refers to the technology, which enables a human to have a real-time sensation of being at a place other than his or her current location. He or she can telexist in a transmitted real world where the robot is located, or in a computer-generated world. Incidentally, telexistence in a computer-generated world is virtual reality in a narrow sense. It is possible to telexist simultaneously in a combination of transmitted and synthesized environments.

Virtual reality in a broad sense is a technology, which invokes a sensation of being present in a realistic virtual environment other than the actual current environment, and provides the means of interacting with the virtual environment in real time (Mann, 1965; Sutherland, 1968; Schmandt, 1983; Brooks, 1986; Fisher *et al.*, 1986). Thus, telexistence and virtual reality are essentially the same technology expressed in different ways.

It was Rheingold (1991) who defined computers as "tools for thoughts." In this sense, the author tends to define virtual reality and/or telexistence as "tools for creation." Furthermore, virtual reality and telexistence are not only tools for creation, but also contain within themselves the possibility of becoming tools, which are useful for various human activities which can be referred to as "human tools for the 3Cs and the 3Es." As mentioned above, the three Cs stand for control, communication, and creation, while the three Es represent experience/education, elucidation, and entertainment.

1.2. Telexistence in the Real World and in Virtual Worlds

Telexistence can be divided into two categories: telexistence in the real world, where the environment actually exists at a certain remote place and is connected via a robot to the place where the user is located, and telexistence in a virtual world which does not actually exist but is created by a computer (Fig. 1.1).

The former can be referred to as "transmitted reality," and the latter as "synthesized reality." Synthesized reality can be further classified into two categories, namely virtual environment as a model of the real world and virtual environment as a model of an imaginary world.

Combinations of transmitted reality and synthesized reality are also possible, and such mixed environments are of great practical importance. Thus, this can be referred to as virtual existence in order to clarify the importance of the harmonic combination of real and virtual worlds.

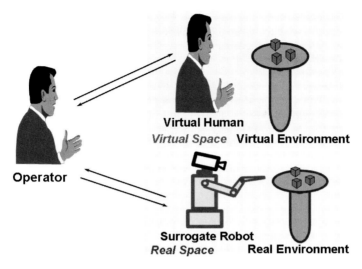

Fig. 1.1.   Telexistence in a real environment and telexistence in a virtual environment.

1.2.1. *Telexistence in the Real World*

Telexistence in the real world, which relies on robots, is further divided into standard and augmented telexistence. The former is intended to facilitate the user's presence in a place which is distant from his or her actual location in real time or with a negligible time delay by means of a robot. The robot is assumed to have the same external shape and sensory functions as a human. In other words, the purpose is to make it possible for the user to perform tasks in a remote environment with the help of a robot. The tasks are performed with the sensation of presence and in real time.

In contrast, augmented telexistence is intended for cases in which the shape/size and/or the sensory functions of the robot are different from those of a human, or the time delay between events in the user's location and the remote location is too long. For instance, augmented telexistence can be applied in cases in which the user controls a microsized or a very large robot.

Although standard telexistence handles only signals in the frequency range of the sensory channels through which the user normally detects stimuli, it is possible to lift such restrictions in order to augment the human sensory functions. This augmentation need not be limited to sensory functions within a human's inherent senses; it can be applied to sensory functions beyond that limit. For this purpose, augmented telexistence employs sensory information obtained through X-rays, ultraviolet rays, infrared rays, microwave radiation, supersonic waves, and ultralow frequency waves sensed by the robot (also referred to as super-sensory information).

For example, if information obtained by a robot, which scans a dark environment for an object by using infrared rays can be displayed by means of computer graphics (CG) and 3D presentation technologies, then the user would be able to see the object on the display screen as if it were brightly illuminated.

Super-sensory information can also be superimposed over the ordinary visual display image through 3D superimposition rather than ordinary 2D superimposition. The possible applications of this functionality include the cases when the distance between the user and the object is displayed in the form of an aerial image, which is superimposed over the location of the object, and when only the portion of an image which has undergone changes for one reason or another is displayed with the added sensation of presence, and the rest of the image originally visible to the user is subtracted.

Augmentation in terms of time is also possible. In the case of the application of telexistence to planetary exploration and other space activities, the delay in the communication time is an important factor for consideration. The maximum permissible time delay, which does not interfere with control in ordinary teleoperation is believed to be about 0.1 s. Anderson and Spong (1989) have proposed an alternative which can keep the system stable with a time delay of up to about 2 s by converting the transmission block into a lossless transmission line.

This makes the transmission block nearly equivalent to a passive element, which appears to be independent of time delay. However, in a system with a time delay greater than 2 s, this method cannot provide proper control.

Even in such cases, augmented telexistence can enable proper control, at least theoretically, if the following method is used with the help of an autonomous remote robot. The robot first scans the environment and prepares an environment model, which is sent to the remote computer system. The system displays visual, auditory, or tactile information with a sensation of presence by taking into account the state of both the object and the user. The user performs tasks in this virtual environment with a sensation of presence, and the essence of the user's work is transmitted to the remote robot.

A robot operating in telexistence mode decodes the information or the instructions for performing tasks and makes additions or applies corrections to the environment model, if necessary. Any discrepancies produced by tasks performed strictly in accordance with the directions imply that there are defects in the model. Therefore, the robot stops operation upon reaching a safe state and then estimates an alternative model. The robot then reports its state and the new model to the user. The user then performs the task again in the virtual environment, which provides a sensation of presence, based on the new model.

Thus, augmentation with respect to time is theoretically possible, although there still remain several unsolved problems, such as the vast number of calculations required and the difficulties associated with the estimation of the model.

1.2.2. *Telexistence in Virtual Worlds*

Virtual worlds created by a computer can be divided into three categories: the physical world, quasi-physical worlds, and non-physical worlds.

In the physical world, the same physical laws apply as those on our planet. Design support and evaluation of virtual products must be performed in the physical world. The virtual environment used for training simulators must also be set in the physical world.

In quasi-physical worlds, the same kinds of physical laws governing the real world apply. However, such worlds can comprise virtual environments corresponding to actual environments such as the Moon, the microscopic world, where the laws of quantum mechanics apply, or a world where the principle of relativity governs the events. The concept of a quasi-physical world is useful for training or educational programs, which provide trainees with experience in a world which is utterly unknown to them.

Considering applications to leisure activities and arts, the potential applications for such fields are not necessarily limited to the physical and quasi-physical worlds. Rather, imaginary worlds are often required in entertainment industries.

Non-physical worlds can meet such demands. Such worlds are technically easier to realize than the former two, although higher artistic sense is often necessary. Artistic expression through virtual reality created in a non-physical world provides a potentially new medium, which embraces linguistic and pictorial means of expression while transcending them at the same time, and can represent ideas and thoughts of sensibility. In this sense, non-physical worlds attract the attention of artists.

1.2.3. *Applications of Telexistence*

Research conducted on virtual reality and telexistence represents an attempt to release humans from spatial restrictions and time constraints.

Based on this perspective, the possible applications of telexistence can be outlined as follows:

(1) Providing substitutes for manual labor in potentially dangerous working environments, such as nuclear facilities, ocean engineering, disaster prevention, and space activities, as well as application in construction work and mining.

(2) Application in secondary industries, i.e. manufacturing industries, using telemachining systems as new production support tools.

(3) Application in primary industries, such as agriculture (telefarmers) and fishing (telefishermen).

(4) Application in tertiary industries including cleaning, maintenance, and other services.

(5) Application in leisure, amusement, and game industries in the form of telexistence surrogate travel.

(6) Application in medical fields, such as microsurgery, telesurgery, and telemedicine.

(7) Application in communication industries, such as communication with a sensation of presence.

(8) Application in the education industry, for example, using experience simulators.

(9) Application in support tools (CAD (computer aided design), IMS (intelligent manufacturing system)) for designing virtual products.

(10) Application in the field of design, including interior design, for developing virtual environments.

(11) Application in scientific-engineering research using virtual scientific visualization as a tool.

(12) Application in research on the behavior of humans and other organisms by using displays providing a sensation of presence.

(13) Providing a new medium of communication which, by embracing linguistic and pictorial expressions and going beyond them, can be used for expressing ideas and sensibility.

1.3. Organization of Telexistence and/or Virtual Reality Systems

The most notable characteristics of virtual reality and/or telexistence is that virtual environments

(1) represent a 3D space which is natural to the user,
(2) allow the user to act freely, and
(3) allow the interaction to take place in a natural form and in real time.

Therefore, the three essential features of virtual reality and/or telexistence are as follows (Fig. 1.2):

(1) providing a 3D space which is natural to the user (3D life-size environment),
(2) allowing the user to act freely and allowing the interaction to take place in a natural form and in real time (real-time interaction), and
(3) providing the user with a realistic projection of himself/herself as a virtual human or a surrogate robot (self projection).

Figure 1.3 shows how telexistence systems are organized. The basic technologies necessary to put telexistence into practice include: (i) the estimation of the user's state (including the external state represented by user movements and tone of voice, as well as the internal state represented by electroencephalograms and electrocardiograms) and the evaluation of the human decision making process; (ii) the interaction between the robot

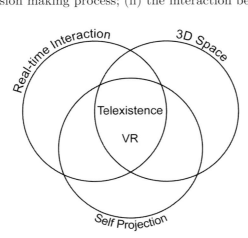

Fig. 1.2. The three fundamental elements of virtual reality and/or telexistence.

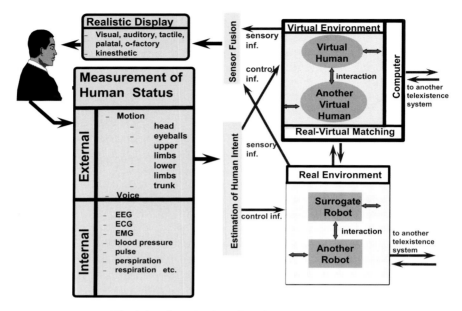

Fig. 1.3.   Organization of a telexistence system.

and the natural environment and/or the interaction between the virtual human and the virtual environment; and (iii) the presentation to the user of the process described in (ii) and its results with the sensation of presence in real time. These are the basic technologies that all the telexistence systems should have in common. An in-depth investigation of such basic technologies is essential to the future development of telexistence. What is characteristic of the study of telexistence is that the achievements of one basic technology are readily available to the others.

In order to prevent virtual reality from becoming a mere application of simulation technology, it is important to connect the virtual and real environments in good harmony. The necessary development of technology necessary for accomplishing this task is one of the problems awaiting a solution.

Furthermore, the following concept provides a new and interesting research subject: a system of a virtual or real environment, which can accept a virtual human or a robot of another telexistence system, presenting them together with the original virtual human or robot. This can give rise to new interactions between virtual humans or robots from two or more systems, in addition to the existing interaction between robots and their environments.

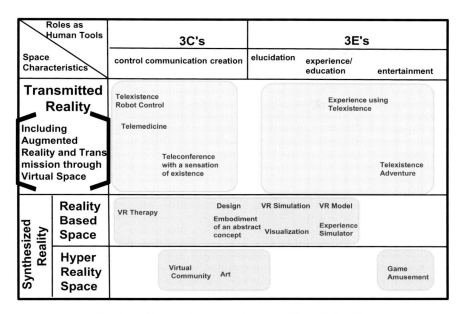

Roles as Human Tools / Space Characteristics	3C's			3E's		
	control	communication	creation	elucidation	experience/ education	entertainment
Transmitted Reality Including Augmented Reality and Transmission through Virtual Space	Telexistence Robot Control Telemedicine Teleconference with a sensation of existence				Experience using Telexistence	Telexistence Adventure
Synthesized Reality — Reality Based Space	VR Therapy	Design Embodiment of an abstract concept Visualization	VR Simulation VR Model Experience Simulator			
Synthesized Reality — Hyper Reality Space	Virtual Community Art					Game Amusement

Fig. 1.4.   VR as a human tool for the 3Cs and the 3Es.

1.4. Virtual Reality as a Human Tool for 3Cs and 3Es

Based on the perspectives mentioned above, virtual reality and/or telexistence provides humans with a tool for the 3Cs and the 3Es, as mentioned before. Figure 1.4 shows the roles of virtual reality as a tool and the characteristics of VR space in the form of a matrix. Typical usages are shown for each combination.

1.4.1. *Control*

In the field of remote manipulation of robots, teleoperation technology, which appeared following the development of nuclear power plants after World War II and the introduction of prostheses for the disabled, developed into supervisory control technology in the 70s. On the other hand, due to the advantages of direct operation, the development of exoskeleton human amplifiers, which cover the human body like armor and protect the body in dangerous environments while increasing the human power at the same time, was also studied in detail. Telexistence is the concept which "aufheben"s the two technologies of supervisory control and exoskeleton human amplifiers, which developed rapidly after 1980.

1.4.2. *Communication*

It is predicted that networked reality can be used not only in offices and factories, but also at home, since the era of the information highway is just around the corner. Figure 1.5 shows an example of the usage of networked reality systems for communication, including the application to communications such as telemeetings, teletravel, teleshopping, and virtual community experiences with a sensation of presence.

This field is closely related to multi-media and Internet, which means that the provision of the network infrastructure is one of the pressing issues (DeFanti *et al.*, 2003). However, what we must keep in mind is that a mere quantitative improvement of the service is not at all sufficient for the dramatic changes in the field of conventional media. In order to make networked media a success, it is necessary to introduce changes in the quality and to introduce technology, which makes it possible to realize concepts such as virtual reality, which were previously considered

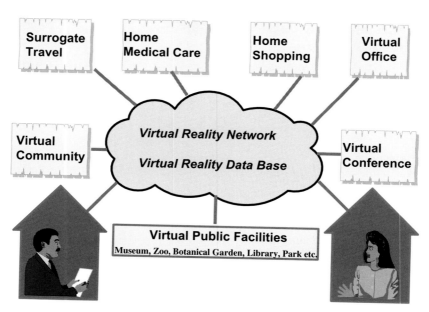

Fig. 1.5.   Conceptual diagram of a networked reality system.

impossible. Thus, virtual reality is expected to be a great incentive to deliver high-speed and high-capacity fiber network access to each home.

1.4.3. *Creation*

Developments with respect to virtual reality can be seen in the fields of design and production. One of the important directions of development of future production systems is to manufacture products, which are adapted to the needs and the individuality of each user without losing exchangeability, expandability, or openness. However, it is not feasible to pursue products which are most appropriate to the needs of each person since the constant repetition of the design and production processes incurs high cost and wastes time, natural resources, and energy.

Virtual reality, which allows computers to be used in the fabrication of virtual products, which evoke the sensation of sight, hearing, and touch in the same way as real products, is expected to solve these problems. This technology also offers useful tools for the amplification of human creativity. We can embody very abstract ideas and concepts in our mind as concrete objects in virtual space and present these objects in a very concrete form to other persons by using virtual reality.

1.4.4. *Experience/Education*

Experience plays an important role in learning, training, and education. Application in education includes, for example, an ultimate simulation including an electronic experience simulator (van Dam *et al.*, 2003; Satava, 2005; Welch *et al.*, 2005).

1.4.5. *Elucidation*

Virtual reality provides rather useful and powerful tools for the elucidation of natural phenomena, for instance, scientific visualization as a tool for scientific-engineering research. Also, the use of VR as a tool for research of the functions of humans and other living creatures is promising.

1.4.6. *Entertainment*

It is not necessary to give a lengthy introduction to this usage since many relevant products are now available on the market. The only due addition is that virtual reality provides a new medium, which embraces linguistic and pictorial means of expression and goes beyond them in expressing human ideas and concepts. Application of virtual reality as human communication media (Naemura, 2005) and artistic expression using virtual reality (Kuma, 2005) are typical examples of this new trend.

1.5. Virtual Reality Convergence

One of the reasons why virtual reality or telexistence attracts worldwide attention is that scientific fields, which have been regarded as belonging to completely different fields of research are likely to be united by the concept of virtual reality, as shown in Fig. 1.6. In the case of remotely controlled robots, the developments of nuclear technologies and orthotic techniques for manufacturing medical braces for handicapped people (such as artificial limbs) were combined after World War II to give rise to teleoperation technology.

In turn, this technology evolved into supervisory control in the 1970s, and, through the adoption of robotics, further developed into telerobotics in the 1980s. As a result, the idea of telexistence-based remote control,

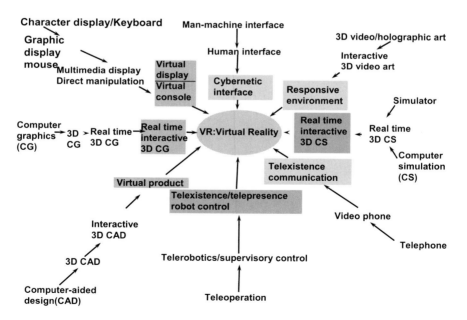

Fig. 1.6. Evolution and development of virtual reality.

which relies on telexistence for a higher degree of sensation of presence on a real-time basis, rapidly evolved later in the 1980s. Thus, we can clearly see these fields of robotics approaching the world of virtual reality.

In the field of CG, the conventional 2.5D display system, in which solid models are shaded and then displayed by using perspective transformation techniques, has advanced to 3D displays which provide the user with stereopsis (stereoscopic vision). This is now developing into an interactive 3D display system in which the image can be changed according to the user's viewpoint. This allows him or her to look sideways or at a downward/upward angle at the image on the display screen as a virtual hologram. The interactive 3D CG techniques which are currently under development are at the shortest distance from the world of virtual reality.

In the field of CAD, attempts are being made to realize a design support/evaluation system based on virtual products by combining CG, tactile sensation feedback, and force sensation feedback. Virtual products can allow designers to evaluate utilization prior to manufacturing and to implement design changes with ease if necessary. Design change data stored in the computer memory is readily available for producing "real" products if the memory is linked to a CIM (computer integrated manufacturing). The concept of virtual reality is of increasing importance to industrial production since it can aid the production of easy-to-use or much more advanced products, which are expected to be in great demand in the future and are more suited to individual user preferences. It also has the merit that it does not waste materials and energy in the production of intermediate products.

In the computer field, more user-friendly human–computer interfaces are desired. In addition to the currently prevalent character input by the use of keyboards, graphical displays, mouse input and object-oriented programming, there are many other possible interfaces, including multi-media displays, 3D mice, and direct input/output operations for communication with a more realistic sensation of presence.

Furthermore, the field of computer-generated simulations is experiencing a rapid development of real-time interactive 3D computerized simulation systems, which are intended for real-time operation on a near-real-experience basis. Simulators are also increasing their reality factor. Flight simulators and surgical simulators are typical examples of this development toward experience-type simulators with a sensation of presence.

The art and amusement industries are no exception. Artists and amusement designers are viewing virtual reality as a new art medium, which could surpass the existing ones with respect to the power and versatility of artistic expression.

In conventional human–machine interface design, humans need to adapt themselves to machines since humans are more flexible than machines. However, the creation of more human-friendly human–machine interfaces has been recently advocated. The next step is a cybernetic interface in which machines can come unilaterally closer to man's natural senses.

The above-mentioned developments will eventually result in the emergence of virtual reality. The rapid progresses in computer and sensor technologies and the increasing number of findings regarding human senses brought about by advances in human science have made virtual reality possible. Recently, many fields of scientific research which have advanced independently of each other have begun to focus on the concepts of virtual reality and telexistence and to view these concepts as key technologies of this 21st century. This encourages firms and organizations with enterprises in such fields to promote intensive research and development programs related to virtual reality and telexistence.

Furthermore, the concepts of virtual reality and telexistence are not simply common to the foregoing fields. The concepts themselves are based on common elemental technology, as will be described in Chap. 4. Therefore, a basic technology developed in one field can be readily applied in another field. This makes it more important to study all related fields as a single generic technology.

Chapter 2

Generations and Design Philosophies of Robots

The short history of robots includes the creation of first- through fourth-generation robots as well as two different philosophies regarding robot design, namely designing robots as independent beings versus designing robots as alter egos.

2.1. Generations of Robots

Since the latter half of the 1960s, robots have been brought from the world of fiction to the real world, and the development of robots is characterized by generations, as in the case of the computer. With the rapid progress and the advent of science and technology after World War II, robots, which had been only a dream for a long time, acquired some manipulatory functions characteristic to humans or animals, although they generally had a drastically different shape. Versatran and Unimate were the first robots made commercially available around 1960. They are referred to as "industrial robots" and can be regarded as the first generation of robots finding practical use.

This is considered to have resulted from a combination of two large areas of development after World War II: hardware configuration and control technology for mechanical hands (or manipulators) for remote operation, which had been under research and development for use in hot radioactive cells in nuclear reactors, as well as automation technology for automated machinery or NC machine tools.

The term "industrial robot" is said to have originated under the patent title "Programmed Article Transfer," which G. C. Devol in the US filed for registration in 1954 and which was registered in 1961 as a patent there. It came into wide use after American Mental Market, a US journal, used the expression in 1960. After passing through infancy in the latter half of the 1960s, the industrial robot reached the age of practical application in the 1970s.

Thus, the robot entered an age of prevalence in anticipation of a rapid increase in demand. This is why 1980 is called "the first year of the prevalence of the industrial robot." From a technical point of view, however, the first-generation robots, i.e., playback robots, were in fact only repetition machines which repeatedly played back their position and posture instructed as an embedded process before commencement of operation.

In essence, first-generation robots were composite systems of technology based on control techniques for various automated machines and NC machine tools, which were combined with design and control techniques and manipulators with multiple (typically 6) degrees of freedom. Naturally, the application area of such robots was somewhat limited to manufacturing processes in secondary industry, especially in material handling, painting, spot welding, and so forth.

However, in certain areas, such as arc welding and assembly, it is necessary to adjust the actions and to better understand human instructions by using not only intrinsic knowledge, as in the case of first-generation robots, but also by acquiring external information with the aid of sensors. A device which can change its actions in accordance with the situation by using sensors is categorized as a second-generation sensor-based adaptive robot. Second-generation robots gradually became prevalent in the 1970s.

The non-manufacturing areas of primary industry (agriculture, fishery, and forestry), secondary industry (mining and construction), and tertiary industry (commerce, service, security, and inspection) have been excluded from the mechanization and automation process, as the older type first- and second-generation robots could not operate in dangerous, unregulated, or unstructured environments. However, harsh and hazardous environments, such as nuclear power plants, deep ocean trenches, and areas affected by natural disasters, are exactly the areas where robots are needed most as substitutes for humans, who risk their lives by working there. Therefore, third-generation robots were proposed to address these problems.

The key to the development of the third-generation robots was to devise a way to enable the robot to work in unmaintained or unstructured environments. First- and second-generation robots acquire data regarding the maintained environment where humans have access to the entire scope of data concerning the environment. This is referred to as "structured environment." Factories where first- and second-generation robots work are typical examples of structured environments. All information concerning the structure of the factory, such as the location of passages and the arrangement of objects, are very clear, and the environment can even be changed to accommodate the robot. For example, objects can be rearranged to locations where the robot's sensors can recognize them easily.

However, there are several types of structured environments which cannot be altered easily. For example, it is not possible to change the environment in places such as the reactors of a nuclear power plant, objects constructed in the ocean, and areas affected by disasters. In such cases, even with full knowledge about the environment, it is difficult or impossible to alter it to accommodate robots. In many cases, the vantage points and the lighting cannot be determined. Furthermore, there are also "unstructured environments" on which humans do not possess accurate data. Many environments exist in nature where humans are completely disoriented.

In the development of third-generation robots, the focus was on the structurization of the environment based on available information. Robots conducted their work automatically once the environment was structured, and worked under the direction of humans in unstructured environments. Such systems, referred to as supervisory controlled autonomous mobile robot systems, were the major paradigms of third-generation robots.

Thus, third-generation robots were capable of working in environments, which were difficult or impossible to alter despite being known to humans from basic data. Such robots engage in security maintenance in such uncontrollable environments, and can handle unpredictable events with the help of humans.

In Japan, for instance, between 1983 and 1991, the Ministry of International Trade and Industry (now Ministry of Economy, Trade and Industry), promoted the research and development of a national large-scale project under this paradigm, which was referred to as "Advanced Robot Technology in Hazardous Environments." The concept of telexistence played an important role in the paradigm of third-generation robots.

Third-generation robots began to work outside factories. However, the environments of such robots are usually restricted to ones where robots can work under the guidance of humans. Although there might be humans working together with the robots, they are usually professionals such as operators or plant workers who have knowledge on robots and/or robotics.

Fourth-generation robots usually operate outside factories, that is, on the street, in hospitals, in offices, or at home, where humans without knowledge of robotics or robots are working together with the robots. Thus, a paradigm shift is essential in this fourth generation.

From the point of view of human society, safety intelligence, alter ego, and anti-anonymity are the three pillars of development of robots coexisting with humans. Safety intelligence requires high technology, and its innovation is not an easy task. Intelligence must be virtually perfect, as only partially successful safety intelligence would be useless. Robots need to possess safety intelligence which matches or even exceeds human intelligence.

As Turing (1950) has argued in his paper, the argument based on the idea that machines think as humans do was not at all relevant in the 1950s or later in the 20th century when autonomous robots were incapable of acquiring true intelligence comparable to that of humans. However, he predicted that it would be relevant to argue about this issue at the end of 20th century. Indeed, as Turing predicted, inventing the safety intelligence is the most important mission in the 21st century, as robots are about to enter the everyday lives of humans.

One could argue that the "alter-ego robots: one's other-self robots" rather than the "independent robots" should be the priority in development. Alter-ego robots are analogous to automobiles, and robots of this type are regarded as machines and tools used by humans, or as extensions of humans in both intellectual and physical sense. For example, a patient can care for himself not only by using his own alter-ego robot, but also by asking family members and professional nurses to take care of him by using robots. Such people, who might live far away from the patient, can use telexistence technology and utilize the robot near the patient in order to help him. One important consideration in using this technology is that the patient needs to perceive the robot as a close person who is taking care of him or her, rather than as an impersonal robot. It is essential that the robot has a "face" as a clear marker of who personifies it.

The analogy with an automobile is effective when considering the importance of clarifying who is using the robot. Just as the driver of a car, as opposed to the car itself, is responsible for the behavior of the car, the person using the robot is responsible for the robot's conduct. When humans interact with a robot controlled by someone else, they need to know who is controlling the robot. The author refers to this concept as the robot's "anti-anonymity." Put simply, the concept refers to the visibility of the face and the body of the robot's user.

Thus, "alter ego," "anti-anonymity," and "safety intelligence" will be the three essential pillars of technical elements in the development of robots in the future.

Robots can be applied in virtually any field imaginable, ranging from home household use to space exploration. However, in the case of fourth-generation of robots, we focus on the humanitarian use of robots, that is, humanitarian robotics, which promotes human welfare activities such as saving human lives. This field of humanitarian use of robots includes robotic surgery, robotic care, rehabilitation robotics, rescue robotics, and robotics for humanitarian demining. Robot technology used in this field can be divided into three parts, namely robot mechanisms, operation cockpits (including human interfaces), and communication technology.

Advanced teleoperation or telexistence plays an important role in the realization of humanitarian robot systems together with technologies such as nano/microtechnology and network technology.

Figure 2.1 summarizes the above discussion regarding the generations of robots.

2.2. Design Philosophies in Robotics

There are two major styles or ways of thinking in designing robots. An important point to note is that these ways of thinking are completely unrelated to the forms of the robots, and there is no distinction between humanoid robots or those with special forms. Other irrelevant distinctions include the possession of general versus specific functions, and those regarding the shape of the robot from the point of view of whether or not it resembles an animal. Nevertheless, these distinctions are indeed important, especially when the robots are put to practical use, and must be considered in practical situations.

However, the distinction discussed here concerns the philosophy of robot design *per se*. The two different ways of thinking concern the question of whether to make "robots as independent beings" or "robots as extensions of humans." Robots as independent beings will ultimately have a will of their own, although this is far from the current stage of development. In this scenario, commands toward the robots are issued by using "language," such as spoken words, written manuals, or computer instruction languages.

On the other hand, robots as extensions of humans cannot have a will of their own. In this case, robots are regarded as parts of the humans who command them, and humans are the only ones who possess will. Commands

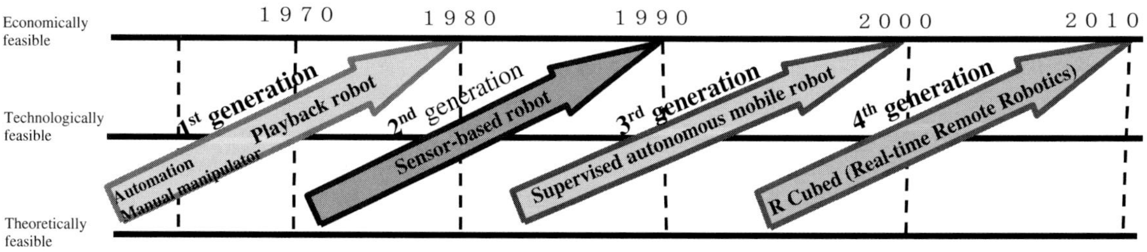

Manual manipulator
Automation manipulator
1st generation Playback robot
2nd generation Sensor-based robot
3rd generation Supervised autonomous mobile robot
4th generation Remote Robotics
R Cubed (Real-time Remote Robotics)

1970   1980   1990   2000   2010   2020

Economically feasible
Technologically feasible
Theoretically feasible

		1st generation	2nd generation	3rd generation	4th generation
Brain Function	Intelligence	A Priori [inborn] Play back	Adaptation Accommodation	Inference Problem Solving	A Posteriori [acquired] Learning, Neural Network
	Knowledge	Data	Data Base	Knowledge Base	Common Sense
Sensor Function	Internal Information	Exist	Exist	Exist	Exist
	External Information	None or Point	1-D, 2-D Structured Environment	3-D Unstructured Environment	3-D Natural Environment
	Communication	Unilateral (teaching, NC tape)	Interactive (robot language)	Bilateral Communication (supervisory control, telexistence)	Bilateral Communication (natural language, gesture, body language)
Effector Function	Manipulation	Position Control (Static)	Position Control (Dynamic)	Force Control	Cooperative Control
	Locomotion	1-D(guide cable)	2-D(plane, guided)	3-D(structured)	3-D (unstructured)
Application Fields		Secondary Industry (manufacturing) Material handling, Painting, Spot welding	Secondary Industry (manufacturing) Arc welding, Assembly	Secondary Industry (nonmanufacturing) Primary Industry Tertiary Industry (inspection, maintenance)	Secondary Industry (nonmanufacturing) Primary Industry Tertiary Industry (office, hospital, home, service)
Technical Feature		Internal Sensor + Servo Control Technology	External Sensor + Microprocessor System Technology	Knowledge Engineering + Man-Machine Interface Technology	Virtual Reality + Network Technology

Fig. 2.1.   Generations of robots.

are issued automatically in accordance with the movements and the internal state of the operator instead of by using language, and robots move in accordance with the human will.

A prime example of robots as extensions of humans is a prosthetic upper limb or an artificial arm which substitutes a lost arm, where humans can move artificial arms as though they move their own arms.

What if it were possible for humans to have an artificial arm as a third arm, in addition to the existing two arms? The artificial arm would move in accordance with the human will and would function as an additional arm extending the human capabilities. An artificial arm, or a robot as an extension of a human, can be physically separated from the human body, while still moving in accordance with the human will, without receiving verbal commands. In such cases, the robot would not have its own will and would function as part of the human, even though the robot is physically separated from the human body. This is what can be referred to as an "alter-ego robot," or "another-self robot." There can be multiple alter-ego robots.

It is also possible to create an environment where humans feel as if they are inside alter-ego robots, thereby allowing the human to survey the environment through the sense organs of the robot and to operate the robot by using its manipulators. This technology is known as telexistence. Telexistence enables humans to transcend time and space, and allows them to be virtually ubiquitous.

Robots as independent beings must have an intelligence that preempts any attempt of robots to harm humans. That is to say, "safety intelligence" is the number one priority in this type of robot design. Isaac Asimov's three laws of robotics, for example, are quite relevant in designing such robots. It is crucial to find a solution in order to make sure that machines would never harm humans by any means.

On the other hand, there is an alternate approach to this problem. One could argue that alter-ego robots or other-self robots, rather than independent robots, should be the priority of development. Alter-ego robots are analogous to automobiles, as pointed out previously. Robots are machines and tools which can be used by humans, and represent extensions of humans in both intellectual and physical sense.

This approach preempts the problem of robots having their own will, as they remain little more than extensions of humans. Humans, therefore, need not feel threatened by robots, as in the case of R.U.R. (Rossum's Universal Robots), which was a play written by the Czech playwright Karel Capek, since the robots remain subordinate to humans at all times. Alter-ego robots are therefore a promising path which humans can follow.

Taking nursing as an example, it is not desirable for a nursing robot which takes care of humans to be an independent being. The privacy of the patients can be protected best when the patients take care of themselves by using the robot. Accordingly, it is more appropriate for the nursing robot to be an alter-ego robot, that is, an extension of a human. Other-self robots can help either their operator or other people. Alter-ego robots are also more secure than robots as independent beings, since the rights and the responsibilities associated with the robot are evident in the former type, where they belong to the humans who operate the robot as their other self. Robots cannot claim their own rights or responsibilities.

In general, there are two major uses for alter-ego robots: one is to transcend time and space by expanding one's existence and the other is to supplement and extend human abilities by using robots as parts or extensions of human bodies, as exemplified by artificial arms and seeing-eye dog robots known as MELDOG (Tachi *et al.*, 1981b, 1985; Tachi and Komoriya, 1985).

Furthermore, the ultimate use of other-self robots would be for humans to copy themselves, including their intelligence. The robot would gradually memorize the intelligence and the behavioral patterns of its user by being used through telexistence, eventually becoming an effective replica of its user.

Computers can imitate language and memory. However, only robots can imitate human behavior. An alter-ego robot spends time with its user as his or her other self or companion, thereby recording the behavior of that human individual. The robot can continue its existence after the user's death, serving as a remnant of the deceased person together with his or her photos and any recorded voice, videotaped images, writings, paintings, and music compositions that the person has left behind. In this sense, alter-ego robots therefore have a more extended use than robots as independent beings.

Thus, "alter-ego," "anti-anonymity," and "safety intelligence" will be the three pillars of technical elements which are essential in the development of future robots. As previously noted, the other self must be part of the user, even though it is physically separated, as the user and its other self share a single consciousness. Accordingly, "other-selfness" is the key to developing robots working in daily life space. Research on this topic must take the concept of the human body into consideration.

Anti-anonymity refers to the visibility of the user's face and body. The robot must indicate who its user is every time it is in use through telexistence via a network. It is unnecessary to discuss how dangerous the world would be without a system which enables the visibility of the user. In addition, the importance of safety intelligence is also evident for other-self robots.

A paradigm shift is essential in this research in the sense that research on intelligence should not be focused on studying and replicating the inner workings of human intelligence. Safety intelligence is a far more important and urgent topic for research. Pursuit of research concerning these three elements is essential for future technology as a whole, and will lead to the realization of 21st century robotics, namely next-generation human–robot networked systems.

Chapter 3

Telexistence

Essentially telexistence is a concept which refers to technology which allows a human to have a real-time sensation of being present in a place other than where he or she is actually located and to interact with the remote environment, which can be real, virtual, or a combination of both. It also refers to an advanced type of teleoperation system, which enables an operator at the controls to perform remote tasks dexterously with the sensation of being inside a surrogate robot working in the remote environment. The application of telexistence in real environments through a virtual environment is also possible.

3.1. Short History of Telexistence

It has long been the desire of humans to be able to project themselves in a remote environment, that is, to have the sensation of being present in a place other than their actual location. Another dream has been to amplify human muscle power and sensing capabilities by using machines while maintaining human dexterity and the sensation of directly performing the task at hand.

In the late 1960s, a research program was planned with the aim of developing a powered exoskeleton which can be worn by the operator as a garment. The concept of a Hardiman exoskeleton was proposed by General Electric Co.; an operator wearing the Hardiman exoskeleton would be able to command a set of mechanical muscles which could multiply his or her strength by a factor of 25, while yet, in this union of human and machine, feeling the object and the applied forces in almost the same manner as in the case of direct contact with the object.

However, the program was unsuccessful due to the following reasons. First, wearing the powered exoskeleton was potentially highly dangerous in the event of the malfunctioning of the machine. And second, autonomous mode was difficult to implement, and everything must be performed by a human operator. Thus, the design proved to be impractical in its original form.

However, with the advent of science and technology, it has become possible to realize this dream by following a different concept. The concept of projecting ourselves by using robots, computers, and a cybernetic human interface is referred to as telexistence. This concept is extended to include projection in a computer-generated virtual environment. Figure 3.1 illustrates the emergence and evolution of the concept of telexistence.

The concept of telexistence was proposed by the author in 1980, and it was the fundamental principle behind the eight-year large-scale Japanese project entitled "Advanced Robot Technology in Hazardous Environments," which was established in 1983 together with the concept of third-generation robotics.

A typical potential use of a third-generation robot is as a substitute for humans in potentially hazardous working environments. These include, for example, work in nuclear power plants, underwater operations, and rescue operations in disaster areas.

One type of robot system which can be used in these areas is a human–robot system consisting of several intelligent mobile robots, a supervisory

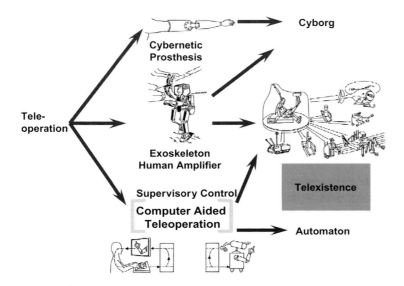

Fig. 3.1. Emergence and evolution of telexistence.

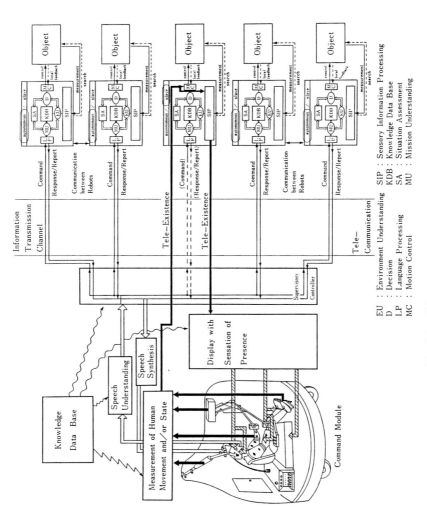

EU : Environment Understanding   SIP : Sensory Information Processing
D : Decision   KDB : Knowledge Data Base
LP : Language Processing   SA : Situation Assessment
MC : Motion Control   MU : Mission Understanding

Fig. 3.2. Telexistence system architecture.

control subsystem, a remote operator, and a communication subsystem linking the operator and the robots. Figure 3.2 shows the author's concept of a human-intelligent robot system, which performs essential work in hazardous working environments.

Planning, scheduling, and task sharing operations for several robots can be handled by a supervisory controller (Sheridan and Ferrell, 1974). Each robot sends its work progress report to the supervisory controller in a consecutive manner. The reports are compiled and processed by the supervisory controller, and selected information is transmitted to the human operator through visual, auditory, and tactile channels. The operator issues macro commands to each of the robots via a voice recognition device.

When an intelligent robot is confronted with a task which is beyond its capabilities, the control mode is switched to a highly advanced type of teleoperation, namely remote-presence, telepresence (Minsky, 1980; Akin *et al.*, 1983) or telexistence (Tachi *et al.*, 1980; Tachi and Abe, 1982; Tachi and Komoriya, 1982). Similar ideas emerged independently in the US and Japan at the same time.

The author referred to this concept of advanced teleoperation as telexistence (Tachi *et al.*, 1980, 1981a, 1984; Tachi and Abe, 1982; Tachi and Komoriya, 1982). Telexistence attempts to enable a human operator at the controls to perform remote manipulation tasks dexterously with the feeling that he or she is present inside a remote anthropomorphic robot in a remote environment.

As mentioned before, fundamental studies regarding the realization of telexistence systems were conducted under the national large-scale project "Advanced Robot Technology in Hazardous Environments," which was a research and development program launched with the purpose of developing a system which eliminates the need for humans to work in potentially hazardous working environments, such as nuclear power plants, underwater, and disaster areas. Through this project, theoretical considerations regarding telexistence have been taken into account, and its systematic design procedure has been established. Experimental hardware telexistence systems have been constructed and the feasibility of the concept demonstrated.

Our first report proposed the principle of a telexistence sensory display and explicitly defined its design procedure. The feasibility of a visual display providing a sensation of presence was demonstrated through psychophysical measurements performed by using an experimental visual telexistence apparatus. Figure 3.3 illustrates the first prototype telexistence

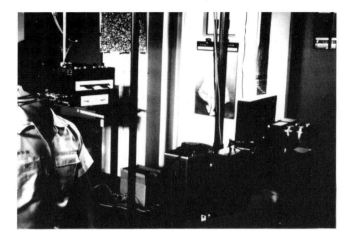

Fig. 3.3.   First prototype telexistence visual display.

visual display (Tachi and Abe, 1982; Tachi and Komoriya, 1982; Tachi *et al.*, 1984).

In 1985, a method was also proposed for the development of a mobile telexistence system, which can be driven remotely with both an auditory and a visual sensation of presence. A prototype mobile televehicle system was constructed, and the feasibility of the method was evaluated (Tachi *et al.*, 1988b).

In 1989, a preliminary evaluation experiment of telexistence was conducted with the first prototype telexistence master–slave system for remote manipulation. An experimental telexistence system for real and/or virtual environments was designed and developed, and the efficacy and superiority of the telexistence master–slave system over conventional master–slave systems was demonstrated experimentally (Tachi *et al.*, 1989, 1990, 1991a,b).

In addition, the first prototype telexistence master–slave system for performing remote manipulation experiments was designed and developed, and a preliminary evaluation experiment of telexistence was subsequently conducted (Tachi and Yasuda, 1994). The slave robot employed an impedance control mechanism for contact tasks and compensated for errors which remained even after performing calibration.

An experimental operation of block-building was then successfully conducted by using a humanoid robot called Telesar (TELExistence Surrogate Anthropomorphic Robot) (Fig. 3.4). Experimental studies of the tracking

Fig. 3.4.   Telesar (TELExistence Surrogate Anthropomorphic Robot).

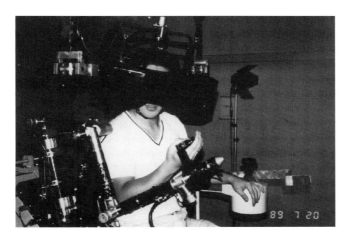

Fig. 3.5.   Telexistence master system.

tasks quantitatively demonstrated that a human can telexist in a remote environment by using a dedicated telexistence master–slave system (Tachi and Yasuda, 1994). Figure 3.5 illustrates a telexistence master system.

3.2. Augmented Telexistence

Telexistence can be divided into two categories: telexistence in a real remote environment, which is linked via a robot to the place where the operator is located, and telexistence in a virtual environment, which does not actually exist but is created by a computer. The former can be referred to as "transmitted reality," and the latter, as "synthesized reality," as was explained in Sec. 1.2.

Synthesized reality can be further classified as a virtual environment representing either a realistic world or an imaginary world. Combining transmitted reality with synthesized reality, which is referred to as augmented reality, is also possible, and it has a great potential for real application. This concept is referred to as augmented telexistence in order to clarify the importance of a harmonic combination of real and virtual worlds.

Augmented telexistence can be used in a variety of situations, such as controlling a slave robot in an environment with poor visibility. An experimental augmented telexistence system was constructed by using augmented reality.

An environment model was also constructed from the design data of the real environment. When augmented reality is used for controlling a slave robot, the environment model must be calibrated in order to eliminate errors. A model-based calibration system using image measurements was proposed for matching the real environment with a virtual environment.

Experimental operation in an environment with poor visibility was successfully conducted by using Telesar (Fig. 3.4) and its dual anthropomorphic virtual telexistence robot named Virtual Telesar, which is shown in Fig. 3.6. Figure 3.7 shows a fragment of an experiment of augmented telexistence (Oyama *et al.*, 1993; Yanagida and Tachi, 1993).

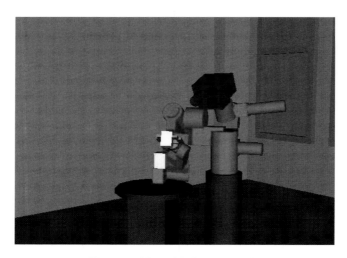

Fig. 3.6.   Virtual Telesar at work.

Fig. 3.7. Superimposition of a model onto a real object.

A quantitative evaluation of the telexistence manipulation system was conducted through tracking tasks by using the telexistence master–slave system described above. Through these experimental studies, it was demonstrated that a human can telexist in a remote environment and/or a computer-generated environment by using a dedicated telexistence system.

Through these research and development programs, it has become possible to implement telexistence between places with dedicated transmission links, such as optical fiber communication links, as was demonstrated by the above experiments. However, it is still difficult for everyone to telexist freely through commercial networks, such as the Internet or next-generation worldwide networks, and more research efforts are anticipated in this area.

3.3. R-Cubed

In order to realize a society where everyone can freely telexist anywhere through a network, the Japanese Ministry of International Trade and Industry (MITI) and the University of Tokyo proposed a national long-term research and development scheme entitled R-Cubed (R^3), which was established in 1995. R^3 stands for real-time remote robotics. The aim of this scheme was the research and development of technologies which enable human operators to telexist freely by integrating robots, virtual reality, and

Fig. 3.8.   Conceptual image of mountain climbing using an R-Cubed robot.

network technology (Tachi, 1998). Figures 3.8 and 3.9 present examples of a conceptual networked telexistence application using R^3 robots.

Figure 3.10 illustrates an example of an R^3 robot system. Each robot site has a server for its local robot. The type of robot varies from a humanoid (high-end) to a movable camera (low-end). A virtual robot can also be a locally controlled system.

Each client has its teleoperation system. It can be either a control cockpit with master manipulators and a head-mounted display (HMD) or a CAVE Automatic Virtual Environment (CAVE) for high-end robots. It is also possible to use an ordinary personal computer system as a control system for low-end robots. In order to assist operators in remotely controlling low-end robots through networks, RCML/RCTP (R-Cubed Manipulation Language/R-Cubed Transfer Protocol) has been developed.

An operator uses a web browser to access the web site describing the robot in the form of hypertext and icon graphics. Clicking on an icon downloads a description file written in the RCML format onto the operator's computer and launches the RCML browser. The RCML browser parses the downloaded file to process the geometrical information,

Fig. 3.9.   Conceptual image of a rescue operation using an R-Cubed robot.

including the arrangement of the degrees of freedom of the robot, the controllable parameters, the available motion ranges, sensor information, and other pertinent information (Fig. 3.11). The browser decides the type and number of devices required to control the remote robot. It then generates a graphical user interface (GUI) panel to control the robot, a video window which displays the images as "seen" by the robot, and a monitor window which lets the operators observe the robot's status from outside the robot. The operator can employ devices such as 6-degree-of-freedom (DOF) position/orientation sensors instead of the conventional GUI panel to indicate the endpoint of the robot's manipulator (see Fig. 3.12).

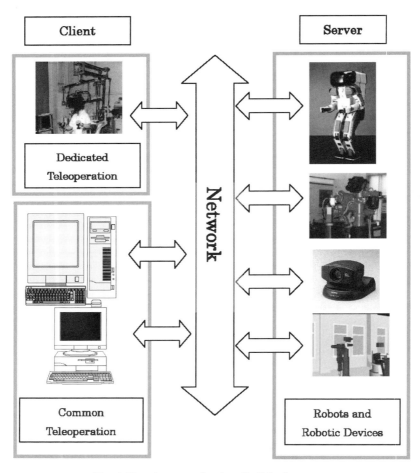

Fig. 3.10.   An example of an R-Cubed system.

3.4. Humanoid Robotics Project: HRP

Based on the R^3 scheme and after a two-year feasibility study entitled
Friendly Network Robotics (FNR), which was conducted between April
1996 and March 1998, the National Applied Science & Technology Project
"Humanoid and Human Friendly Robotics," or "Humanoid Robotics
Project (HRP)" in short, was launched in 1998. This five-year project aimed
at developing a safe and reliable human-friendly robot system capable of
carrying out complicated tasks and supporting humans within the sphere of
human lives and activities. One of the goals of the project was research and

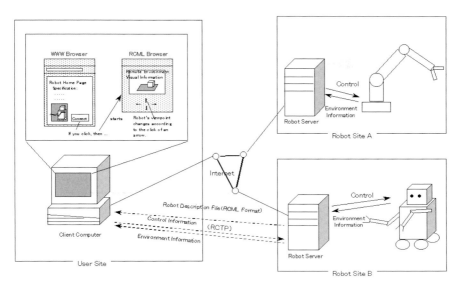

Fig. 3.11.　Diagram for RCML and RCTP process.

Fig. 3.12.　RCML browser.

development oriented toward the realization of a so-called R-Cubed society by devising humanoids, control cockpits, and remote control protocols.

The outline of the project, which was announced by NEDO (New Energy and Industrial Technology Development Organization) at the beginning of the project, was as follows:

Japan's population is aging rapidly and people are having fewer children. This means that efficient and human-friendly machinery that can support daily life and the activities of humans, such as attending to the elderly and the handicapped, is in great demand. Thus, this project aims to develop a safe and reliable human friendly robot system capable of carrying out complicated tasks and supporting humans within the sphere of human lives and activities. Through development of such a system, this project will contribute towards improvement in efficiency and safety in industry, making society and the living environment more convenient and comfortable and towards the creation of new industries in the manufacturing and service sectors.

This project will be undertaken in two phases. The goal of phase one is to develop a platform (basic research models) for human-friendly and supportive robot systems by bringing together the elemental technologies possessed by industry, academia and the government with the latest technologies. To be specific, R&D will be carried out on the following themes:

(1) *Development of a platform for a human-friendly and supportive robot;*

(2) *Development of a virtual platform for a human-friendly and supportive robot.*

   In phase two, R&D will be carried out towards the application of human-friendly and supportive robots with consideration given to the needs of industries in which such robots might be used. Improvement and addition of elemental technologies will be carried out using the platform and the virtual platform developed in phase one. R&D will be carried out on the following theme:

(3) *Application of a platform for a human-friendly and supportive robot.*

Figure 3.13 shows how HRP was carried out. In phase 1 of the project (1998–2000), a humanoid robot was constructed by Honda, while

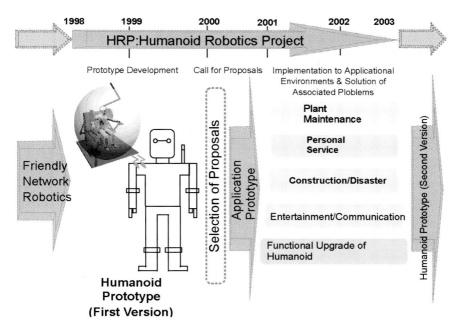

Fig. 3.13.   HRP (Humanoid Robotics Project) general plan.

a telexistence cockpit was developed by Matsushita Electric Works (now Panasonic Electric Works), Kawasaki Heavy Industries, FANAC, and the University of Tokyo as a platform for a human-friendly and supportive robot. The virtual robot platform was developed by AIST (National Institute of Advanced Industrial Science and Technology), the University of Tokyo, Waseda University, Fujitsu, and Hitachi.

In phase 2 of the project (2000–2003), HRP2 was developed by Kawada Industries. Several operations were carried out by Mitsubishi Heavy Industries, Kawasaki Heavy Industries, Matsushita Electric Works, Hitachi, Fujitsu, Yasakawa, Tokyu Construction Co., Shimizu Corporation, ALSOK, the University of Tokyo, Kyoto University, Tohoku University, Osaka University, Tokyo Institute of Technology, University of Tsukuba, Hiroshima City University, Waseda University, and AIST.

It is necessary to use a novel robot system capable of assisting and cooperating with people for activities such as the maintenance of nuclear plants or power stations, construction work, supply of aid in case of emergencies or disasters, and care for elderly people. If considerations are taken into account from both technical and safety points of view, it is clearly intractable to develop a completely autonomous robot system for these objectives.

Robot systems should therefore be realized by using a combination of autonomous control and teleoperated control. By introducing telexistence techniques through an advanced type of teleoperated robot system, a human operator can be provided with information regarding the robot's remote environment in the form of natural audio, visual, and tactile feedback, thus providing the operator with a sensation of being present inside the robot itself.

Thus, in phase 1 of the project, a telexistence cockpit for controlling humanoid robots was developed (Fig. 3.14), and the telexistence system was constructed by using the developed humanoid platform.

In order to address the problem of narrow fields of view associated with conventional HMDs, a surround visual display using immersive projection technology similar to that adopted in the CAVE (Cruz-Neira *et al.*, 1993) was developed. The surround visual display presents a panoramic view composed of images captured with a stereo multi-camera system mounted on the robot for a wide field of view, which allows the operator to feel the on-board motion when he or she uses the robot to navigate the remote environment.

Fig. 3.14.   Telexistence cockpit for controlling humanoid robots.

When using the teleoperation master system, a human operator sits on a seat at the motion base, attaches the gripping operation device, and grips the master arm. Through the master arm and the gripping operation device, the operator can remotely manipulate the robot arms with his or her hands. The motion base can display vibration, shock, and acceleration acting on the robot, as well as the relative displacement of the upper part of the robot's body from a reference position based on the inclination of the operator.

Each master arm is designed as an exoskeleton-type arm and has seven DOF. This redundancy in the DOF allows the operator to place the slave arm directly in various positions by using his or her elbow, whose motion is tracked by a joint motor on each master arm and measured by optical sensors located on the lower links. The remaining joint motors generate appropriate force (up to $10\,N$) on the basis of the feedback force from the slave arm, and therefore the operator feels force and moment naturally.

Each master arm has a gripping device which the operator can use to easily produce open–close motion by feeling the gripping force of the slave robot. In order to realize small and lightweight mechanisms as well as a wide operation space for the thumb and index finger, a wire tension mechanism with passive DOF is used to facilitate the thumb's radial abduction and ulnar adduction.

The developed motion base system allows the operator to experience the locomotion of the humanoid robot in a realistic manner by representing its acceleration, posture, and motion. The motion base provides the operator with a sensation of walking, up-and-down stepping, and body inclination by manipulating the seat position under the operator's standing posture. In order to minimize the displacement of the operator's focal point, the motion base system limits locomotive motion to a 3-DOF translation: back and forth, left and right, and up and down.

Audio and visual information is transmitted through an analogue communication module. The control information is exchanged and shared through a shared memory module referred to as a reflective memory module, which is shared and accessed by the audio/visual display subsystem, the teleoperation master subsystem, and the control system of the robot itself.

Various teleoperation tests were carried out by using the developed teleoperation master system, and the results show that kinesthetic presentation by using the master system with visual input greatly improves the operator's walking experience and dexterity with regard to manipulating objects. If the operator issues a command to move the robot, the robot actually walks toward the goal. As the robot navigates its environment, real images captured by the multi-camera system with a wide field of view are displayed on the four screens of the surround visual display. This provides the operator with a feeling as if he or she is inside the robot in the remote environment.

In order to evaluate the usability of the cockpit for HRP robots, an experiment was conducted where an operator was given the task to navigate the environment of the robot and manipulate objects by utilizing the humanoid robot as a surrogate. For the purpose of demonstrating the feasibility of using the developed system in the field of service robots, a mockup shopping zone was built in a real environment 3.5 by 6.0 m in size. A humanoid robot was set inside the mockup, and a human operator was given the task to control the robot from the telexistence cockpit in the remote site with a sensation of presence. The operator navigated the robot as if he or she were inside the robot, and manipulated the robot's arms and hands to handle a stuffed animal, stack blocks, open and close a sliding glass door of a cupboard, pick up a can from inside the cupboard, place the can into a basket, and so forth.

When the operator controlled the robot to walk, he or she wore polarized glasses and leaned on the sheet of the motion base. On the bottom left screen of the surround visual display, there was an operational menu which included a 2D map of the environment and a series of operational commands to the robot. The operator used a 3D mouse to indicate on the map the location and orientation of the target which the robot had to reach, and the menu system automatically generated a path to reach the goal. If the operator issued a command to move the robot, the robot actually walked toward the goal. While the robot walked, real images captured by the multi-camera system for the wide field of view were displayed on four screens of the surround visual display. This provided the operator with a feeling as if he or she were inside the robot in the remote environment.

In addition, a CG model of the robot in the virtual environment was represented and updated according to the current location and orientation received from the real robot. It was displayed on the bottom right screen of the surround visual display, and when it was added to the real images captured by the camera system, it supported the operator's navigation of the robot.

Figure 3.15 presents the robot descending stairs. Since the series of real images presented on the visual display are integrated with the movement of

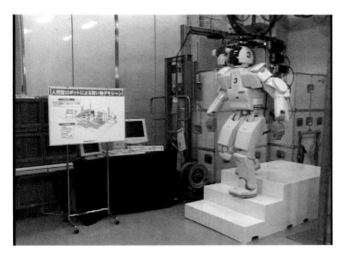

Fig. 3.15.   HRP humanoid robot at work.

the motion base, the operator experiences a real-time sensation of walking or stepping up and down.

This was the first experiment in the world which succeeded in controlling a humanoid biped robot by using telexistence (Tachi *et al.*, 2001, 2003).

Chapter 4

Fundamental Technologies for Telexistence

4.1. Telexistence Visual Display (Tachi and Abe, 1982; Tachi and Komoriya, 1982; Tachi *et al.*, 1984)

It is desirable for an operator at the remote control console to have a real-time sensation of presence as if he or she were inside of a remote anthropomorphic robot and to be able to maneuver it dexterously. This concept is referred to as telexistence. The realization of a visual display which provides a sensation of presence is one of the most important elements of this telexistence. In this section, a method for realizing a telexistence visual display and its design procedure is proposed and explicitly defined. An experimental visual display system was assembled, and the feasibility of the visual display with a sensation of presence was demonstrated by performing psychophysical experiments using the test hardware. The validity of the proposed design method was demonstrated after comparing the Helmholtz horopters from observations using a telexistence system and direct observations.

4.1.1. *Design Concept of Telexistence Visual Display*

Figure 4.1 gives the configuration of a basic telexistence system, and a concrete method for the configuration of a visual display providing a sensation of presence is explained. Figure 4.1(a) shows the principle of the recording and reproduction of wave fronts in a holographic manner, as conceived in the past. In other words, a closed surface is created so as to surround the remote place, and the wave fronts entering that area are recorded at multiple points on the enclosure surface.

These wave fronts are then transmitted to the location of the local observer and are subsequently reconstructed by using a reproduction device on a similar enclosure surface surrounding the observer. However, this

(a)

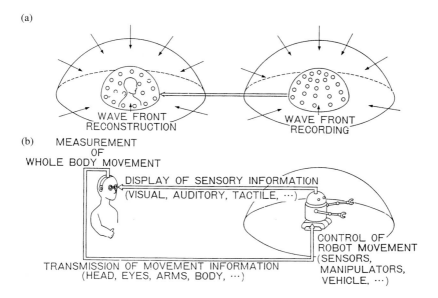

Fig. 4.1.   Principle of telexistence.

method alone makes it difficult to realize telexistence for the following reasons:

(1) If the recording and reproduction device is designed so as to reconstruct the actual environment, it would be too large to be practical. In addition, in holography, the recording and reproduction of real-time information cannot be achieved with the presently available technology.

(2) Although the display of a distant background scene can be approximated by a large 2D screen, technically, it is extremely difficult to achieve a 3D reproduction of nearby objects in actual size and in real time without them being occluded by real objects.

(3) In particular, the actual sensation of presence cannot be achieved if the hands of the operator are located at a place different from those of the robot. In telexistence, the hands of the robot must be visible at the place where those of the human operator ought to appear. However, the realization of such a state is generally difficult when using the above-mentioned holographic method.

As a consequence, it was not possible to utilize this conventional method to acquire a true sensation of presence, i.e. the sensation of being inside the robot or at the place where the robot is located, which would be produced by the realization of the appropriate relationship between the background scene, the task at hand, and the hands of the robot.

Figure 4.1(b) shows the proposed method for configuring telexistence on the basis of robot technology and the human sensory structure. According to the method described in Fig. 4.1(a), an attempt is made to reproduce all wave fronts at the same time. However, this is unnecessary in view of the working mechanism of the human visual perception system. The basis of human visual perception is a pair of images focused on the retinas, and the wave fronts perceived by a person as retinal images at any given instant of time constitute only part of the total wave fronts. These wave fronts change in real time with the movements of the person's head and eyes. The person then creates, inside his or her brain, a 3D world based on two images which change with time, and projects it back to the place where the objects really exist.

As a result, if it is possible to measure the movements of a person's head and eyes faithfully and in real time, to move the head and the eyes of the robot in line with those movements, to transmit to the human side the pair of images created in the visual input device of the robot at that time, and to recreate these images on the person's retinas accurately and without a time delay by using a suitable display device, then that person can receive retinal images equivalent to those seen directly by the robot. In other words, by using these images, the person can create inside his or her brain a 3D world equivalent to the one seen directly from the place where the robot is located, and then project it again into the real world.

Thus, by scanning partial wave fronts continuously by using a human motion measurement device and a system consisting of a display device and a slave robot, a recording and reproduction device can be produced which is small enough to be realistically configured, and therefore problem (1) presented above can be solved.

In addition, this method creates a state in which visual information of the direct observation on the human side is shielded, and instead, visual information from the remote robot is displayed as if the operator is at the same place where the robot is located. Moreover, as the movements of the human's hands, arms, and torso are measured faithfully, and this information is used in order to move the manipulators and the body of the robot, when the operator moves his or her own hands and arms in front of his or her eyes, a configuration becomes possible in which the robot's manipulators appear in front of its eyes at the same position where his or her hands and arms are supposed to be.

As a result, problems (2) and (3), which are relevant in the case of conventional display systems, such as the one in Fig. 4.1(a), can also be solved.

4.1.2. *Design Method and Procedures of*
Telexistence Visual Display

Figure 4.2 shows the method of configuring an ideal telexistence visual display system. Two cameras are placed at the same distance as that between the eyes of a person. The remote head mechanism on which these cameras are mounted is controlled in conjunction with the movement of the person's head, as stated previously.

First, the movements of a person's eyes are measured, as illustrated in Fig. 4.1(a), and in conjunction with this information, the vergence angle θ_r of the cameras and the vergence angle θ_h of the video display (a CRT, LCD, LED, or OEL display) are adjusted so as to obtain $\theta_r = \theta_h$. At the same time, X_r is determined, so that the cameras are focused correctly. Furthermore, the lens system, which is situated in front of the video display, is adjusted so as to obtain $X_h = X_r$ for the position of the virtual image of the video display and $I_h = I_r$ for the size of the image, respectively.

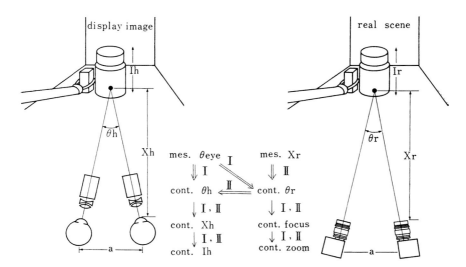

Fig. 4.2.   Ideal visual display method of a telexistence system.

There are three parameters which need to be taken into consideration when a person perceives distance: (1) the accommodation of the crystalline lens, (2) the size of retinal images, and (3) the vergence angle of the two eyes. In the event that the ideal system in Fig. 4.2 is used for creating a realistic perception in the person using the system, the values of parameters (1)–(3) above are equal to those of parameters (1)–(3) in direct observation, respectively. As a consequence, the ideal system can assure the reproduction of the same visual information as that in direct observation.

Next, an attempt is made to simplify the ideal system. An investigation of the features of human visual perception makes it clear that regarding (1), if the crystalline lenses are fixed at 200 mm, the natural fusion of the images takes place in a vergence range between 100 and 500 mm, and if they are fixed at 1,000 mm, the fusion corresponds to a vergence range between 200 mm and infinity. As a result, the sensation of presence is not lost even if the distance X_h in the image displayed in Fig. 4.2 is constant at $X_h = 1,000$ mm. This fact was verified during an experiment on spatial perception.

If the display distance is fixed at 1,000 mm, then there remain only two variables which need to be controlled: the size of the retinal images and the vergence angle of the eyes. Controlling two parameters instead of three clearly simplifies the ideal system. The actual design method based on the simplified method of composition is presented in Fig. 4.3.

Figure 4.3(a) shows the characteristics of direct observation performed by a human. The distance between the person's eyes is denoted by W_m, and the distance from the center of the crystalline lenses to the retinas is denoted by a_m.

Now, if by observing an object with a size ℓ_{obj} located at a distance d_{obj} from the observer, the vergence angle α and the size ℓ_m of the image of the object on the retinas can be obtained, after which the distance to the object and its size can be determined by using the known parameters and the observation values as follows:

$$d_{\text{obj}} = \frac{W_m}{2} \cot\left(\frac{\alpha}{2}\right), \tag{4.1a}$$

$$\ell_{\text{obj}} = d_{\text{obj}} \cdot \frac{\ell_m}{a_m}. \tag{4.1b}$$

A plane vertical to the observation direction located at a distance d_{vir} in front of the observer is considered in place of the actual image. Furthermore, by carrying out the perspective transformation of the observed object, with

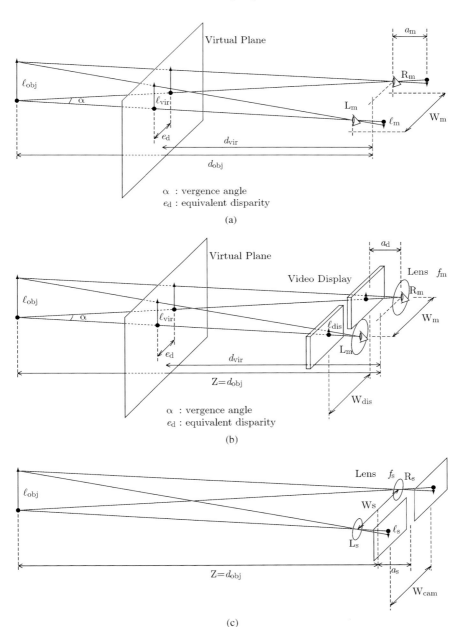

Fig. 4.3.   Visual parameters of (a) direct observation, (b) a telexistence visual display, and (c) a slave robot camera system.

its projection center at the center of the crystalline lenses, its size on that virtual plane is denoted by ℓ_{vir}, and the shift of the left and right transformed images on the virtual plane (equivalent disparity) is denoted by e_d.

Using the symbols ℓ_{vir} and e_d, Eq. (4.1) can now be rewritten as follows:

$$d_{\mathrm{obj}} = \frac{W_m \cdot d_{\mathrm{vir}}}{(W_m - e_d)}, \tag{4.2a}$$

$$\ell_{\mathrm{obj}} = d_{\mathrm{obj}} \cdot \frac{l_{\mathrm{vir}}}{d_{\mathrm{vir}}}. \tag{4.2b}$$

In other words, instead of placing an object of size ℓ_{obj} at a distance d_{obj}, a perspective transformation image ℓ_{vir} of ℓ_{obj} is placed at e_d on the virtual plane located at a distance d_{vir}. If at this time the distance d_{vir} is fixed at 1,000 mm so as to be able to ignore the effects of the muscular tension and the relaxation of the crystalline lenses, then it is possible to obtain the same effect as that for the time when the object is actually observed.

Figure 4.3(b) presents a method for realizing the configuration in Fig. 4.3(a). Figure 4.3(c) shows the configuration of the video camera system on the side of the slave robot used for obtaining the image displayed on the video display at that time. The distance W_s between the lenses of the two video cameras on the slave side is set equal to the distance W_m between the eyes of the observer (master).

$$W_s = W_m. \tag{4.3}$$

The distance W_{cam} between the two video sensing elements (such as CCD, MOS, etc.) of the video cameras and the central distance W_{dis} between a pair of video displays are set equal to W_s.

Even though they have different physical dimensions, it is possible to satisfy the conditions of Eq. (4.4) equivalently by using an optical method or an electric method involving scanning line processing, etc.

$$W_{\mathrm{cam}} = W_{\mathrm{dis}} = W_s. \tag{4.4}$$

Under these conditions, the image enlargement ratios β (as a function of d_{obj}) and γ are defined as follows:

$$\beta \equiv \left\lfloor \frac{\ell_{\mathrm{dis}}}{\ell_{\mathrm{obj}}} \right\rfloor_{\text{at } d_{\mathrm{obj}}}, \tag{4.5a}$$

$$\gamma \equiv \frac{\ell_{\mathrm{dis}}}{\ell_s}, \tag{4.5b}$$

where ℓ_{dis} is the size of the image on the video display and ℓ_s is the size of the image on the imaging plane of the camera.

Now that the distance between the imaging plane and the lens of the camera, which has a lens of a focal distance f_s, is given by $a_s = f_s \cdot d_{\text{obj}}/(d_{\text{obj}} - f_s) \approx f_s$, the video display must be placed at location a_d in front of the observer so as to satisfy Eq. (4.6)

$$a_d = \beta \cdot d_{\text{obj}}, \tag{4.6a}$$

$$a_d = \gamma \cdot a_s. \tag{4.6b}$$

Finally, the images of the video display must be moved onto the virtual plane by placing a convex lens at a focal distance f_m in front of the eyes of the observer. In other words, f_m is used, which satisfies Eq. (4.7)

$$f_m = \frac{a_d \cdot d_{\text{vir}}}{d_{\text{vir}} - a_d}. \tag{4.7}$$

4.1.3. *Visual Display Prototypes*

4.1.3.1. *Visual Display Unit*

Two types of color display units were prepared on a trial basis. One was prepared with a special emphasis on image quality by using 4-inch color CRTs. The other was composed of 1.5-inch color CRTs and was designed to be lightweight. Each of these units used MOS-type semiconductor color cameras as their input cameras. It was possible to set the focal distance arbitrarily between 12.5 and 75 mm with a computer command. The iris diaphragm was adjusted automatically. Figure 4.4 shows the 4-inch display configuration, while Fig. 4.5 illustrates the 1.5-inch display configuration. The values of β for $f_s = 12.5$ mm are 0.16 and 0.053, respectively. In addition, the values of γ are 12.9 and 4.25, respectively. As a result, the optimum parameters were determined by setting $a_d = 161$ mm and $f_m = 192$ mm for the 4-inch system and $a_d = 53$ mm and $f_m = 56$ mm for the 1.5-inch system. Furthermore, in the case of the 1.5-inch system, its vergence angle was controlled with signals issued from the computer. Incidentally, the weights of the displays of the 4-inch and 1.5-inch systems were about 10 kgf and 6 kgf, respectively.

4.1.3.2. *Experimental Visual Display System*

Figure 4.6 shows a schematic diagram of the experimental system. The movements of the person's head are measured by using a goniometer with six degrees of freedom (DOFs), which is fixed on the helmet, and these

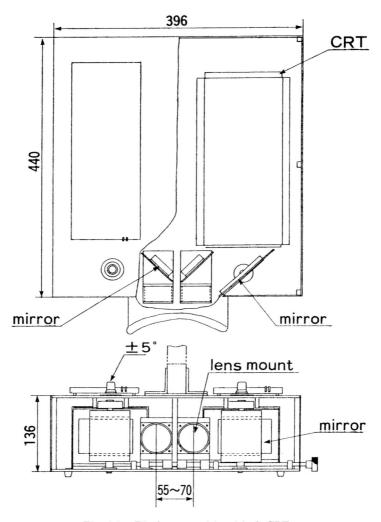

Fig. 4.4.   Display comprising 4-inch CRTs.

movements are subsequently subjected to coordinate transformation in the computer. On the basis of this information, the active visual display unit, which also has six DOFs, is coordinated with the movements of the person. At the same time, the input mechanism comprising a pair of cameras with five DOFs is also coordinated with the movements of the person.

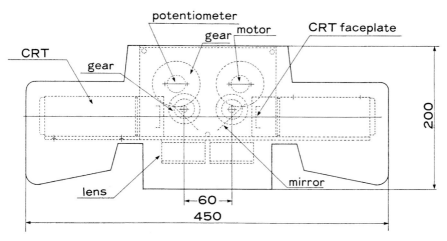

Fig. 4.5. Display comprising 1.5-inch CRTs with a servo-controlled vergence mechanism.

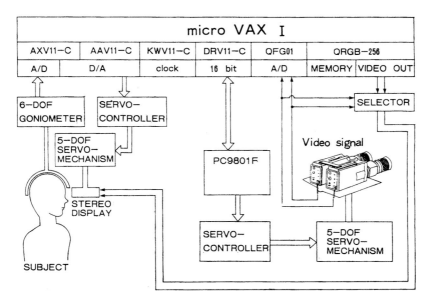

Fig. 4.6. Schematic diagram of the experimental system.

The multiple-DOF display system is composed of a goniometer with six DOFs for measuring the movements of the head and a master–slave type display device for minimizing the feeling of head constraint.

One of the problems arising when the DOF of the head movement is increased is the feeling of constraint due to the weight of the display unit. An experiment has already been carried out with one DOF, in which the feeling of constraint due to the inertia of the display unit was an issue. The 1-DOF experimental apparatus was configured in such a way as to allow rotation of the unified body of the goniometer and the display unit in horizontal direction by using the power of the neck muscles. Although the weight of the display device (about 4 kgf) was supported on a rotating axle, considerable power was required for rotating and stopping the display device. It was feared that the sensation of presence would be degraded due to the feeling of constraint caused by the band in the head section.

Here, since measurements are prepared with respect to rotations not only to the left or to the right, but also to back and forth movements, it is necessary to move the display device up and down in conjunction with these movements. If the force of inertia and the gravitational force acting on the display device need to be supported by the neck, the burden on the neck increases further due to the increase of the weight of the display system, which was designed to display color. Under these circumstances, the test device created here has a separate display device and a goniometer with an actuator on the side of the display device, so that the burden on the neck comes only from the goniometer. This is a type of master–slave system in which the display device moves in conjunction with the movement of the head, which is measured with the goniometer in real time.

However, since the goniometer and the arm of the display device must be positioned in such a way as to overlap almost completely, it is difficult to make the shapes of the goniometer and the display device identical. In addition, the position and orientation of a single point on the head section as measured with the goniometer are not necessarily identical to the position and orientation of the display device. For this reason, it is necessary to calculate the coordinates which place the detector on the master side and the actuator on the slave side in a one-to-one correspondence.

A general view of the apparatus is presented in Fig. 4.7. The goniometer has very low frictional resistance (less than 10 gf) in horizontal direction since it comprises a horizontal two-link mechanism. The weight of the helmet and the goniometer (1.4 kgf) is compensated by using a spring. A total of six DOFs in terms of position and posture can be measured.

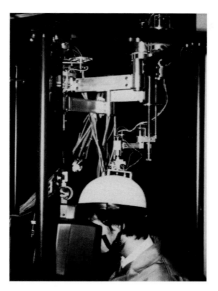

Fig. 4.7. General view of the experimental hardware featuring a 6-DOF goniometer (right) and a 5-DOF servo-controlled binocular display.

The driving mechanism of the display unit has five DOFs. The rotational movement around the line of vision is omitted since it is irrelevant for the field of view. With the use of a 5-joint link, the structure was made symmetric with respect to its left and right sides, and the effect of the gravitational force was easily compensated by using a mechanism which enables movements in horizontal direction to be independent of vertical movements. A direct-drive mechanism was realized by using a torque motor provided by Inland Corp. as an actuator.

4.1.4. *Evaluation of the Visual Display*

4.1.4.1. *Horopters and Their Expression*

The most important point in evaluating a visual display device is that the psychological visual space as seen directly by the viewer should be in a one-to-one correspondence with the psychological visual space as seen through the visual display device. In order to investigate this quantitatively, the psychological visual space is approximated as a Riemann space of constant curvature (Luneburg, 1950), and its parameters are compared with

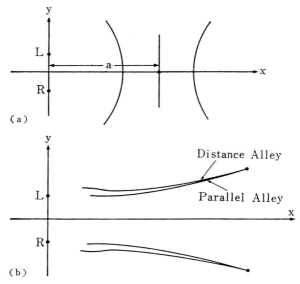

Fig. 4.8.   Helmholtz horopter curves for different fixations (a) and (b) distance and parallel alleys.

respect to the case of observation through the device and the case of direct observation.

As shown in Fig. 4.8(a), a horizontal plane at the height of the eyes of the observer is considered, and the x axis is defined along the forward direction inside that plane. The y axis is in the direction across the face of the observer and perpendicular to the x axis. Here, L and R denote the positions of the left and the right eye of the observer.

The head of the observer is fixed inside a dark room. Multiple small points of illumination (light points) are presented on the horizontal plane in front of the observer.

Next, the observer attempts to rearrange those light points in such a way that they are perceived by the observer as parallel to the facial plane, which is parallel to the y axis. It is not always the case that those points, which are arranged physically in parallel, appear to be parallel to the observer. A physical arrangement which appears to be parallel is seen differently depending on the distance from the observer to the parallel line.

That is to say, although a physically straight arrangement located at a certain distance from the observer is perceived as a straight line parallel to the facial plane, an arrangement on a convex plane relative to the observer is perceived as a straight line from a larger distance. Moreover, at a closer distance, an arrangement on a concave plane relative to the observer is perceived as a straight line. Such curves are referred to as Helmholtz horopters.

Figure 4.8(b) shows a physical arrangement, referred to as Hildebrand's parallel alley, which appears parallel to the x axis. In addition, an arrangement of two points equidistant along the y axis yields a different result from a parallel arrangement, and this arrangement is referred to as Blumenfeld's distance alley.

Such horopters and alleys show the same tendency regardless of who the observer is. Moreover, although the specific forms of the horopters and the alleys depend on the observer, they are always constant as long as only a single person is involved.

Figure 4.9 shows a schematic diagram of the process of mapping of the physical space coordinate system to the psychological visual space. Now, if an investigation is conducted in order to find out whether a difference exists between the transformations in direct observation and those perceived via a display system, such as in a telexistence system, then

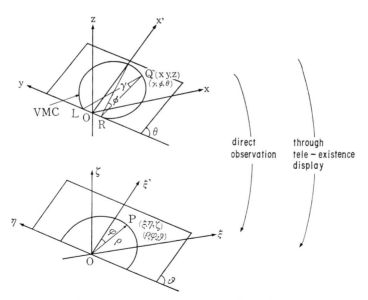

Fig. 4.9.   Mapping of physical space (binocular coordinates) onto visual space (polar coordinates) for direct observation and through telexistence display with various display parameters.

that display system can be evaluated. In other words, by comparing the results of horopters and alleys in direct observations with those obtained in observations via a telexistence display system, the display system can be evaluated quantitatively.

There are two ways of expressing horopters and alleys: over a psychological space, or directly over a physical space.

The respective Euclidean expressions of horopters and alleys in a psychological space can be approximated by using the following equations, after setting the coordinate system of the psychological visual space to (ξ, η, ζ), as shown in Fig. 4.9, and assuming that the variables in the diagram satisfy the relations $\phi = \varphi$, $\theta = \vartheta$, and $\rho = 2\exp(-\sigma\gamma)$

$$\frac{K}{4}(\xi^2 + \eta^2) - 1 = A\xi, \tag{4.8a}$$

$$\frac{K}{4}(\xi^2 + \eta^2) - 1 = B\eta, \tag{4.8b}$$

$$\frac{K}{4}(\xi^2 + \eta^2) + 1 = C\eta. \tag{4.8c}$$

The expressions in Eq. (4.8) are those of horopters in a psychological visual space, and are effective in showing that a visual space is a non-Euclidean space, particularly a hyperbolic space of the Lobachevski and Bolyai type.

However, these expressions are defective in that their respective meanings are difficult to comprehend intuitively, and also that the estimation of the parameters K, A, and σ is unstable in general, unless a fairly large number of points are measured.

Other expressions of horopters are based on a method for describing the horopter curves themselves in a physical space. In general, they can be described as a group of curves satisfying the relation $f(x, y) = 0$.

In this case, an experiment in horopter measurement is carried out as described below. By using three light points, the ones at the two ends are fixed on a straight line parallel to the facial plane, and the one in the center is moved to the location where it appears to be exactly on that straight line.

In other words, with $x = r$ denoting the distance from the facial plane to the measurement point, the coordinates of the two points are fixed as $(r, -y_0)$ and (r, y_0), where y_0 is positive. Next, the x coordinate of the light point at the center is adjusted in such a way that when it is located at point

$(\bar{x}, 0)$, all three points are assumed to be on a straight line. Then,

$$\Delta x = \bar{x} - r, \tag{4.9}$$

where Δx is a function of x, and the horopters can be expressed by using this $\Delta x(x)$. In practice, $x_i (i = 1, \ldots, n)$ are obtained, and $(\bar{x} = \sum_{i=1}^{n} x_i/n)$ is used.

In the present experiment, the horopters are evaluated quantitatively by using Δx from Eq. (4.9).

4.1.4.2. Experiment

By using two small light emitting diodes (LEDs), which are 3 mm in diameter, inside a dark room, small light points are created and arranged so that $y_0 = 150$ mm at points $(r, -y_0)$ and (r, y_0). Here, r can be 1, 1.5, 2, 2.5, or 3 m, and in each case, a third LED is placed at the center denoted by $(x, 0)$.

The central LED is placed on the x–y table, whose position can be controlled so that the operator can adjust the position of the LED freely by controlling the x coordinate.

A subject with his or her head fixed by means of a head rest, measures his or her horopter by adjusting the x coordinate of the central illumination point in order to ensure that the three illumination points are lying on a straight line parallel to the y axis.

Using the same conditions, this experiment is also carried out for the case of direct observation and the case of observation via a display system.

Figure 4.10(a) shows an example of the experiments on horopters through direct observation, while Fig. 4.10(b) presents an example of the results of the horopter experiments conducted by fixing the vergence angle of the display so as to make it equal to the vergence angle of a direct observation at 2 m. The symbols $\bigcirc$ and $\square$ in the central section of the diagram correspond to the measurements. The horopter curves are obtained by connecting the average values of the measurements and the two ends by means of smooth curves which are symmetric with respect to the x axis.

A 1.5-inch system is used in this case. Since $d_{vir} = 2$ m and $a_d = 55$ mm, the optimum parameters for telexistence are given as $f_s = 13$ mm and $f_m = 57$ mm.

Since the vergence angle is adjusted for a distance of 2 m, if there is any shift from that point, the horopters are found to be different from the ones based on direct observation. Figure 4.10(c) compares the horopters

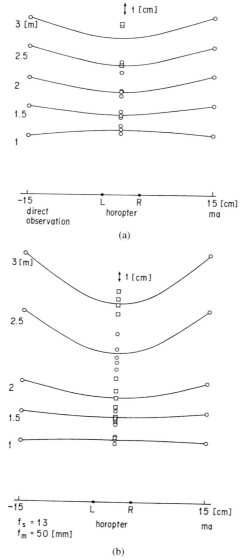

Fig. 4.10. Experimental results for $d_{vir} = 2\,m$ using a 1.5-inch display (subject ma): (a) horopter for direct observation, (b) indirect observation with $f_s = 13\,mm$ and $f_m = 50\,mm$ for a 1.5-inch display, and (c) comparison between different display parameters of a 1.5-inch display at $d_{vir} = 2\,m$.

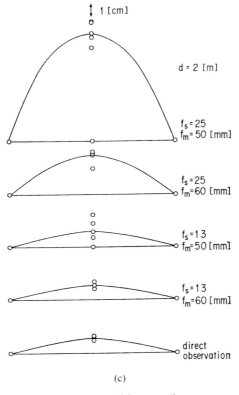

(c)

Fig. 4.10.   (*Continued*)

obtained from an observation distance of 2 m for various display parameters. It is clear that when the focal lengths of the eyepiece and the objective lenses are varied from their proper values, the size of the images on the retinas changes, or when the vergence angle is varied, the horopters deviate from the ones for direct observation. The variance of the data increases accordingly.

The relationship between the focal lengths of the eyepiece and the objective lens for which the horopter pattern shows the best match with the one for direct observation, is the nearest relationship determined by the proposed design method, namely $f_s = 13$ mm and $f_m = 60$ mm. In this case, Δx (at $x = 2$ m) $= 11$ mm is the same as 11 mm for the direct observations. In this way, the proposed design method is shown to be valid. Incidentally, even if f_m is 50 mm, as long as f_s is 13 mm, then the average value is $\Delta x = 12$ mm, which is in nearly complete agreement with the value for direct observation. Therefore, the system is expected to perform well in practice.

Figure 4.11 presents various results for the case in which the observation distance is set to 1.0 m. A set of five displays is given in each image, and their average is indicated with the symbol ◯, where the vertical bar indicates the upper and lower limits, and therefore the range of data variation.

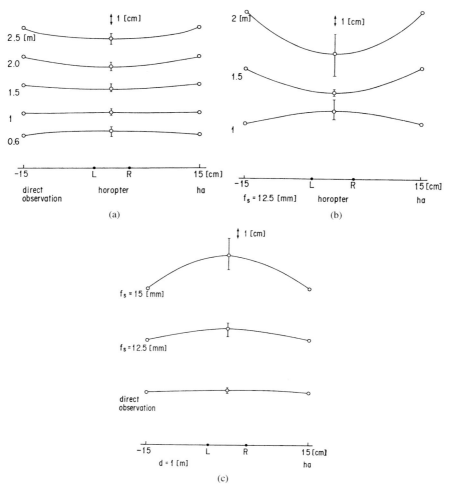

Fig. 4.11. Experimental results for $d_{\text{vir}} = 1$ m using a 4-inch display (subject ha): (a) horopter for direct observation, (b) indirect observation with $f_s = 12.5$ mm and $f_m = 190$ mm for a 4-inch display, and (c) comparison between different display parameters of a 4-inch display at $d_{\text{vir}} = 1$ m.

For this experiment, the 4-inch display was used. Since the optimum display conditions were determined to be $d_{\text{vir}} = 1\,\text{m}$ and $a_d = 262\,\text{m}$, $f_s = 12.5\,\text{m}$ and $f_m = 192\,\text{mm}$. With f_m fixed at 190 mm, f_s was varied from 10 to 12.5, 15, and 20. As a result, the corresponding values for Δx (at $x = 1\,\text{m}$) were -15, 13, 32, and 40 mm, respectively. It is thus clear that the result most closely approached the value for Δx of 3 mm of the horopter in direct observation when the focal length f_s of the lens on the camera side was 12.5 mm. In addition, the data variance was small for the case of 12.5 mm.

It is shown by these experiments on horopters that the proposed telexistence design method is valid in terms of the cognition of space.

It should be pointed out here that although these results were obtained only for a horizontal plane in front of the face of the observer, it was possible to obtain the horopters even when the display plane was off the horizontal plane and to consider them as equal to those for direct observation as long as the angle of elevation was within the range of $\pm 10°$.

Moreover, if the display range is wider, the head of the observer must necessarily move. In the proposed system, the slave camera system faces the same direction as the head of the observer. This means that as long as the observer can see the central section of the field of vision, the same conditions obtained in the experiments with the horizontal plane are maintained for wider view. Thus, the same state as that of direct observation can be ensured for wide display ranges through the telexistence visual display.

4.1.5. *Summary*

A method was proposed to realize a telexistence visual display, and its design procedure was explicitly defined. A psychophysical experimental method was proposed, in which Helmholtz horopters were used as measures of visual space for the purpose of evaluating the difference between directly observed visual space and visual space observed indirectly through the telexistence display system. Experimental telexistence display systems were prepared, and the feasibility of the visual display with a sensation of presence was demonstrated by performing psychophysical experiments using the proposed evaluation method.

4.2. Mobile Telexistence System (Tachi *et al.*, 1988b)

A prototype mobile telexistence system was constructed which enables a human operator at the controls to perform remote tasks dexterously with the feeling that he or she is present at the place in a remote environment where the slave mobile robot is located, and the feasibility of the mobile telexistence was evaluated. The system consists of an independent mobile robot with two TV cameras, a remote control station with visual and auditory displays providing a sensation of presence, and a communication link between the human operator and the mobile robot. The effectiveness of the proposed system was evaluated through navigation experiments involving the navigation of a mobile robot through an obstructed space. Several display and operation methods were compared quantitatively by using the elapsed time, the smoothness of the traveled path, and the number of collisions as the criteria for comparison.

4.2.1. *System Configuration*

A prototype system with fundamental mobile telexistence functions was assembled for experimentation purposes. Figure 4.12 shows a schematic diagram of the experimental system, which consists of an independent mobile robot with two color TV cameras, a remote control station with visual and auditory displays which provide a sensation of presence, and a communication link between the human operator and the mobile robot.

The head movements of the human operator are measured in real time, and the robot's vision system is controlled to follow the movements of the operator. The robot can be navigated either by an autonomous control system or through commands issued by the operator. The images acquired by the robot's vision system are transmitted and displayed to the operator's eyes through an appropriate lens system by using head-linked video displays. The two images in turn fuse to give a very natural visual perception.

The remote station is controlled by a microprocessor (PC9801 VM2 with D-board 16 and microVAX 11). The movements of the operator's head and the position of the control knob and switches are measured in real time, processed by the computer, and sent to the mobile robot via a wireless modem (HD-9600-ACH).

The mobile robot is a battery-operated three-wheeled cart which is controlled by a microprocessor (PC 9801 VM2) with a coprocessor (D-board 16). Command signals are received through an on-board wireless modem (HD-9600/B-ACH), and the processed information is used for controlling the movement of the TV cameras, the direction and the velocity of the propulsion subsystem, the steering subsystem, and the brake subsystem.

Two video signals from the two color TV cameras are transmitted in turn to the remote station via two UHF transmitters (Ch. 22 and Ch. 26).

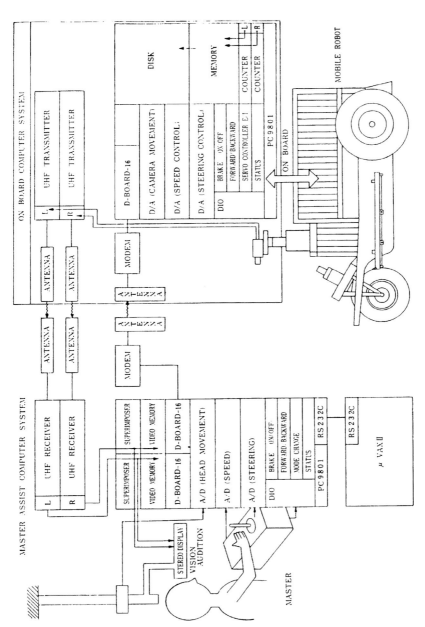

Fig. 4.12. Experimental mobile telexistence system.

Video signals are received through the UHF receivers and are conveyed to the visual and auditory display either directly or through computer superimposers.

Two video displays with appropriate lens systems are placed immediately in front of the operator's eyes. Remote scenes taken by the left and right cameras onboard the robot are displayed on the left and right displays, which are focused by the lens systems on the corresponding left and right retinas, respectively. The visual angle at which each eye sees the object on the video display is controlled so that it coincides with that of the direct observation, and the disparities between the two corresponding pictures on the two displays are controlled so that the distance to the displayed object is always maintained the same as that of the real object. This system was designed in accordance with the design procedure described in Sec. 4.1.

Auditory information is perceived through a headset. The left and right signals are received from microphones attached to the left and right TV cameras, respectively.

During routine navigation tasks, the robot travels autonomously using the environmental map and the environmental information gathered by the visual sensors (two TV cameras and an ultrasonic sensor) and internal sensors (two odometers on the rear wheels). Visual information is processed remotely by the microVAX 11, while ultrasonic and odometer signals are processed by the microprocessor onboard the robot.

The navigation process can be monitored by the operator. When it encounters a task which it is not capable of managing by itself, the robot stops and asks the operator for help. At that time, the operator controls the robot by using joysticks in a manner similar to driving an automobile, i.e. as if the operator was onboard the robot and his or her eyes were at the position where the robot's TV cameras are located.

Figure 4.13 shows the prototype telexistence mobile robot, and Fig. 4.14 shows the head-linked display providing a sensation of presence used in the system. This telexistence display system is designed on the basis of the method described in Sec. 4.1 using 4-inch CRT displays.

4.2.2. *Experiments*

In order to evaluate the effects of the application of the telexistence display to the mobile robot system described above, a comparison between a conventional display and a telexistence display was conducted.

Fig. 4.13.   Prototype mobile telexistence vehicle (televehicle).

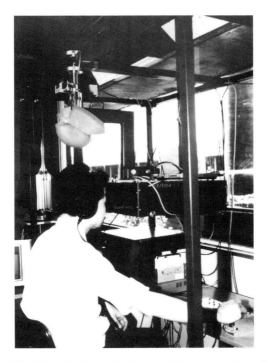

Fig. 4.14.   Head-linked stereo display providing a sensation of presence.

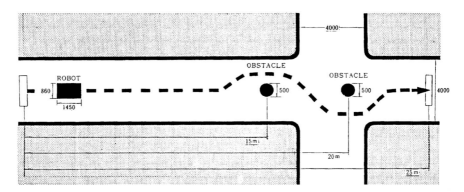

Fig. 4.15.   Experimental conditions.

Figure 4.15 shows the tasks which the remote operator is given for execution. The goal is set at a distance of 25 m, and two cylindrical obstacles block the way. The remote operator controls the mobile robot while observing the situation by using either a conventional 2D display (14-inch TV monitor) or the telexistence display. The operator uses a joystick which controls the velocity and the direction of the mobile robot's motion.

Three operational modes are set for the telexistence display and the conventional display, as shown in Fig. 4.16. In the independent mode, the camera follows the head movement of the human operator, while the steering is controlled by the operator using the joystick. In the follow-steering mode, the camera follows the movement of the steering, which is

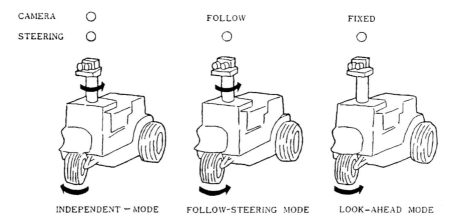

Fig. 4.16.   Three operational modes tested in the experiments.

Table 4.1. Comparison of the time needed for different operational modes.

	Look ahead	Follow steering	Independent
2D Display without reference	Difficult	Difficult	Difficult
2D Display with reference	5′43″	4′35″	4′28″
Telexistence display without reference	3′18″	2′12″	2′6″
Telexistence display with reference	2′40″	1′15″	1′5″

also controlled by the operator using the joystick. In the look-ahead mode, the camera is fixed in the forward direction.

Preliminary experiments revealed that it is very helpful for the remote operator to have a reference which indicates the orientation of the robot's body. Therefore, a rectangular frame with a reference at the center of the front side is fixed in front of the robot (see Fig. 4.13) in order for the operator to perceive the orientation of the robot's body through the display.

Table 4.1 shows the results of the experiments. The table lists the time necessary for one of the five experimental subjects to reach the goal for the combination of the four display types and the three operational modes with a fixed camera angle of 48° (the numbers represent the average of three trials).

For the other subjects, the same tendency was observed, although the absolute values varied from subject to subject. The robot was quite difficult to operate with the conventional TV display, and collisions occurred frequently. Both the independent mode and the follow-steering mode show superior results when using the telexistence display together with the reference. This coincides with the subjective feeling of all five subjects that these are quite natural and easy to operate.

In order to analyze the difference between the independent mode and the follow-steering mode, the trajectories of the mobile robot under those two operational modes were measured and compared. Figure 4.17 shows the best result for the independent mode, while Fig. 4.18 shows the best result for the follow-steering mode. It was confirmed on the basis of Fig. 4.17(a) that the vehicle stopped when the operator surveyed the environment by turning his head in the independent mode. This increased the total elapsed time and made the trajectory somewhat uneven.

In the follow-steering mode presented in Fig. 4.18(a), the televehicle did not stop during surveillance, which reduced the elapsed time and made the trajectory smoother (Fig. 4.18(b)).

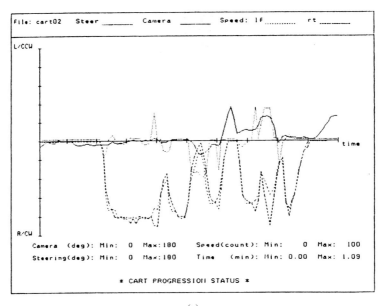

(a)

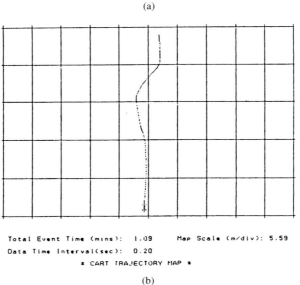

(b)

Fig. 4.17. (a) Recorded wheel speeds, steering angle, and camera angle for the independent operation mode, and (b) estimated trajectories under the independent operation mode.

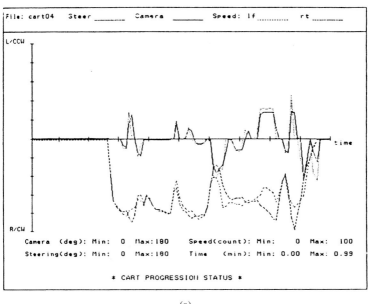

(a)

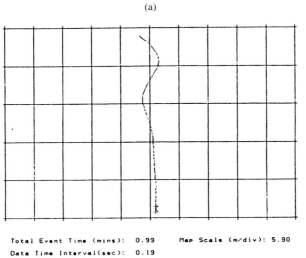

(b)

Fig. 4.18. (a) Recorded wheel speeds, steering angle, and camera angle for the follow-steering mode, and (b) estimated trajectories under the follow-steering mode.

4.2.3. *Summary*

The effectiveness of the proposed telexistence mobile system was evaluated by performing navigation experiments involving the navigation of the mobile robot through an obstructed space. Several display and operation methods were compared quantitatively by using the time needed for accomplishing the task, the smoothness of the traveled path, and the number of collisions as the criteria for comparison.

The follow-steering operational mode used together with the telexistence display and a vehicle orientation reference showed the best results. The independent mode was found to be useful for the operator to obtain an accurate view of the environment by stopping the televehicle.

4.3. Design and Quantitative Evaluation of Telexistence Manipulation System (Tachi and Yasuda, 1994)

In this section, a telexistence manipulation system is evaluated quantitatively by comparing the performance in tracking a randomly moving target under several operational conditions. The effects of various characteristics, e.g. binocular vision and the effect of natural arrangement of the head and the arms, are analyzed by quantitatively comparing the results under these operational conditions. A human tracking transfer function was measured and used for comparison. The results revealed the significant superiority of the binocular vision and the natural arrangement of the head and the arms, which are the most important characteristics of telexistence.

4.3.1. *Telexistence Manipulation System*

Figure 4.19 shows a schematic diagram of the telexistence master–slave manipulation system, which consists of a master system with a head-coupled 3D visual and auditory display, a master manipulator, a computer control system, and an anthropomorphic slave robot mechanism with an arm having seven DOFs, a gripper hand, and a locomotive mechanism.

A human operator wearing a 3D audiovisual display, which is designed to ensure the same distance and size cues as in direct observation from the location of the robot, is shown in Fig. 4.20. The audiovisual display is carried by a link mechanism with six DOFs. The link mechanism cancels all gravitational force through a counterbalancing mechanism, which allows the operator's unconstrained movement in a relatively wide operation space (up/down: -500–400 mm; right/left: -300–300 mm; forward/backward: -300–800 mm).

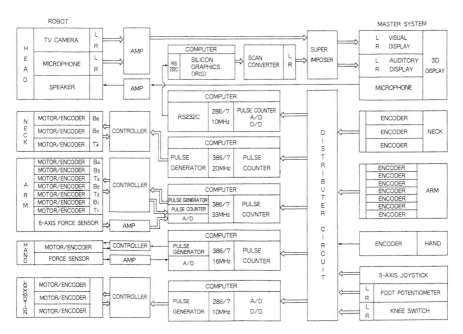

Fig. 4.19. Block diagram of the telexistence master–slave manipulation system.

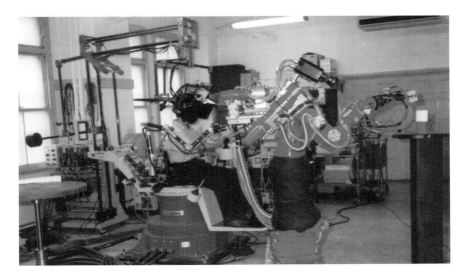

Fig. 4.20. General view of the telexistence master–slave manipulation system.

This setup also enables the display to follow the operator's head movements with sufficient precision to permit ordinary head movement. Regarding rotational movements, the mechanism is designed in such a way that the three rotational axes meet at one point. A parallel link mechanism is also used for attaining the roll motion as well as for load bearing. The DOFs are arranged in such a way that the most important yaw motion (pan) is available at any orientation. The maximum inertial force applied to the head of the operator remains less than 5 kgf (Fig. 4.21).

The master arm has ten DOFs. Seven DOFs are allocated for the arm itself, and three additional degrees are used to comply with the body movements.

The movements of the operator's head, right arm, right hand, as well as other auxiliary movements (including the joystick operation and the feet movements) are measured by the master motion measurement system in real time without constraint. The measured signals corresponding to movements of the head, the arm, the hand, and any auxiliary movements are sent to computers. There are four computers (equipped with Intel 286/386 processors) which generate the command position of the slave head movement, the arm movement, the hand movement, and the locomotion of the slave robot, respectively.

All programs were written in the C language and run under MS-DOS, and the program sizes are 11572, 76983, 22611, and 86375 bytes, respectively. Calculations on each computer are synchronized by the

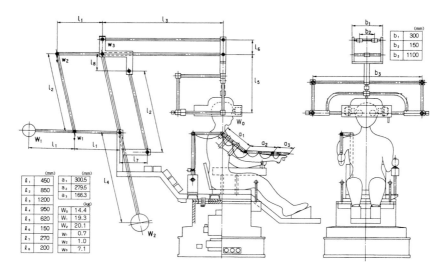

Fig. 4.21.   Telexistence master system.

movements of the human operator through the master system so that all computers are automatically coordinated.

The servo controller controls the movements of the slave anthropomorphic robot. The slave robot has a locomotive mechanism and a hand mechanism. The robot also has three DOFs in the neck mechanism, on which a stereo camera is mounted. It has an arm with seven DOFs, and a torso mechanism with one DOF (waist twist). The dimensions and the arrangement of the DOFs of the robot are designed to mimic those of the human body.

The range of each DOF is set so that it can cover the movements of a human, while the speed is set to match the average speed of human movements ($3\,\text{m/s}$ at the wrist). The weight of the robot is $60\,\text{kgf}$, and the arm can carry a $1\,\text{kgf}$ load at a maximum speed of $3\,\text{m/s}$. The precision of position control of the wrist is $\pm 1\,\text{mm}$. A six-axis force sensor installed at the wrist joint of the slave robot measures the force and the torque exerted upon contact with an object, which is used to adjust the mechanical impedance of the robot's arm to a compliant predetermined value.

The robot moves by a planar motion mechanism whose position is assigned in a polar coordinate system (r, θ), where $r = 500\text{--}1{,}500\,\text{mm}$ and $\theta = 0\text{--}270^\circ$. The orientation of the robot is assigned by the waist rotation angle ϕ of the robot, where $\phi = -150\text{--}150^\circ$. A hand mechanism with one DOF, which can either pinch or grasp, was also designed. It is capable of pinching small objects (with a diameter of $2\,\text{mm}$) as well as rather large objects (with a diameter of up to $114\,\text{mm}$).

The hand uses a parallel link mechanism and a ball screw. Grasping of cylindrical objects with a minimum diameter of $15\,\text{mm}$ can be performed by making contact at three points, which makes the grasping stable. Strain gauges, which measure the grasping force, are placed on two finger links. The average grasping force is $5\,\text{kgf}$. Furthermore, measurements of the opening are performed by using an encoder attached to a DC motor. Position control with an average resolution of $0.01\,\text{mm}$ is attained, and a six-axis force sensor is installed at the wrist position. The hand is made of durable aluminum and weighs $620\,\text{gf}$, including the force sensor.

The vision system of the slave robot consists of two color CCD video heads from TV cameras. Each CCD has a resolution of 420,000 pixels and its own optical system with a focal length of $f = 12\,\text{mm}$ (field of view 40°) and an aperture of F1.6. Focus is automatically controlled with the TTL AF method. The distance between the two cameras is set to $65\,\text{mm}$, and the two cameras are aligned parallel to each other.

Regarding the auditory system, two microphones are placed $243\,\text{mm}$ apart from each other, and the same positional relation is used for the auditory display of the master system. A small speaker is placed at the location of the mouth, which transmits the operator's voice.

The stereo visual and auditory input system mounted on the neck mechanism of the slave robot gathers visual and auditory information from the remote environment.

This information is sent back to the master system and is applied to the specially designed stereo display system in order to evoke the sensation of presence at the remote location in the operator. The measured information on the human movement is used to change the viewing angle, the distance to the object, and the conditions of the object and the hand in real time. The operator observes the 3D remote environment in front of his or her eyes, which changes according to their movements.

The stereo visual display is designed in accordance with a developed procedure which assures that the 3D view maintains the same spatial relations as in the case of direct observation, as described in detail in Sec. 4.1. A pair of 6-inch LCDs (H720 × V240 pixels) are used together with a convex lens system. The compact arrangement of a display system suitable for use as the master manipulation system was made possible by arranging the two mirrors so that the LCDs can be placed on the upper side in front of the operator (Fig. 4.22) (Maeda and Tachi, 1992).

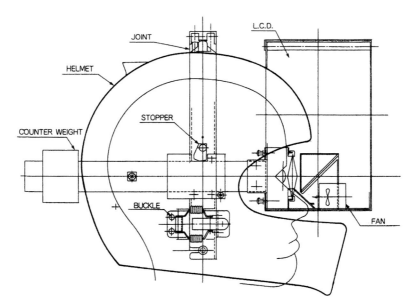

Fig. 4.22.   Telexistence head-mounted display.

4.3.2. *Manipulation Experiments*

Experiments which quantitatively evaluate the typical characteristics of the telexistence master–slave system were conducted. The most noticeable differences between telexistence and/or virtual reality and the conventional human-machine interface is that a virtual environment providing a sensation of presence has the following features, as is described in Chap. 1:

(1) The virtual environment is a 3D space which feels natural to the user (3D Life-sized Space),
(2) it allows the user to act freely and allows the interaction to take place with natural movements in real time (Real-Time Interaction), and
(3) it allows the projection of the operator in the form of a virtual humanoid or a surrogate robot (Self-Projection).

Thus, the most important features of telexistence include natural 3D vision (closely approximating direct observation), which follows the operator's head movements in real time, and a natural correspondence between visual information and kinesthetic information, where the operator observes the slave's anthropomorphic arm at the position where his or her arm should appear. This is regarded as the basis of the feeling of existence, which allows the operator at the control to perform tasks which require real-time coordination of hand and eye, as in the case of direct object manipulation.

In order to obtain experimental proof and a quantitative evaluation of the effects of the three features of telexistence, the following experiment was conducted. The following five visual display methods were compared:

(1) direct observation;
(2) HMD (B): binocular head-mounted display and a stereo camera mounted on the slave robot, whose orientation, i.e. pitch, roll, and yaw, are controlled to follow the movements of the operator;
(3) HMD (M): monocular head-mounted display and one camera mounted on the slave robot, whose orientation, i.e. pitch, roll, and yaw, are controlled to follow the movements of the operator;
(4) CRT (H): a conventional CRT display placed in front of the operator with a field of vision of 45° and a camera placed at the eye position of the robot head, whose orientation is fixed to the direction of the movement of the target;
(5) CRT (O): conventional CRT display placed in front of the operator with a field of vision of 45° and a camera placed 30° to the side of the

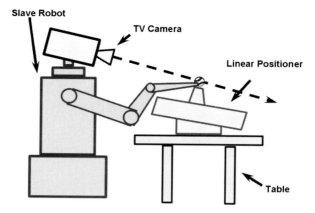

Fig. 4.23. Experimental arrangement.

robot, whose orientation is fixed to the direction of the movement of the target.

The head-mounted display was designed in accordance with the procedure described in Sec. 4.1. In the HMD (M) mode, only the right-side display of the binocular system was used. The field of view was 40° for each eye.

Figure 4.23 shows the experimental arrangement of the slave robot and the linear positioner. A target was fixed to the moving part of the linear positioner, which was driven by random noise with a maximum stroke of ±100 mm along the depth axis of the operator's observation direction.

The operator was asked to place the tip of the slave robot's manipulator at the position of the target by using the master manipulator under several display conditions (conditions 2 through 5). Since the target moved randomly, the operator was forced to follow the target (tracking). In this experiment, the target's moving direction was elaborately arranged so that it coincides with the direction of observation in order to eliminate mutually dependent effects and to single out the effects of the head movements of the operator, the binocular observation, and the matching of kinesthetic and visual information.

Under condition 1 (direct observation), the operator was at the position of the slave robot instead of the robot and tracked the target with the master manipulator while observing the target directly. This condition was used as the control data.

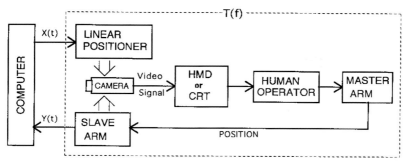

(Sampling Time : 30ms, 1024points)

Fig. 4.24.   Block diagram of the evaluation system.

Pseudorandom noise was used as the target position input, as follows:

$$x(t) = \sum_{k=0}^{n} a_0 p^{-k} \sin(2\pi f_0 p^k t + \phi_k),$$

where $p = 1.25$, $n = 17$, $f_0 = 0.0326\,\text{Hz}$, and ϕ_k is a random number.

The experimental tracking system is shown in Fig. 4.24. As shown in the figure, the pseudorandom noise $x(t)$ was applied to the linear positioner containing the target, and the operator attempted to follow the random movements of the target by manipulating the master manipulator while observing the target and the slave manipulator through a HMD or a conventional CRT display under several display conditions.

The human operator's tracking trajectory $y(t)$ along the linear positioner's coordinate was calculated by using the kinematics of the slave manipulator and the measured seven joint angles of the manipulator. The performance was evaluated by comparing the transfer function of the human operator $T(f)$, which was estimated using $x(t)$ and $y(t)$ for each of the above-mentioned display methods.

$T(f)$ is estimated as follows:

$$T(f) = \frac{\Phi_{xy}}{\Phi_{xx}} = E\left[\frac{X(f)^* Y(f)}{X(f)^* X(f)}\right],$$

where Φ_{xy} is the cross-spectrum between the input signal $x(t)$ and the output signal $y(t)$, and Φ_{xx} is the power spectrum of signal $x(t)$. Signals $x(t)$ and $y(t)$ are measured for a finite amount of time in order to determine their Fourier transforms. The uppercase letters denote the Fourier transform of the corresponding signals denoted by lowercase letters. In addition, the asterisk denotes the complex conjugate, and $E[\ ]$ denotes the ensemble average.

The control cycle of the master–slave system and the linear positioner was 10 ms. The output response was sampled every 30 ms. FFTs (fast Fourier transforms) of 1,024 points were employed, and the cross-spectrum was measured by using the frequency averaging technique for each of the display methods. This process was repeated five times in order to obtain an ensemble average of the cross-spectrum, after which the transfer function was estimated as the ratio of the average cross-spectrum and the power spectrum for each display method.

Figure 4.25 shows an example transfer function. The amplitude (gain) and the phase of the human transfer function $T(f)$ under the tracking task are shown as a function of the frequency. The crossover model can be applied as a first-order approximation. According to the crossover model of McRuer (McRuer and Jex, 1967; Sheridan and Ferrell, 1974), the transfer function $T(f)$ in the region of the crossover frequency can be described as follows:

$$T(f) = (\omega_c/j\omega)\exp\{-j\omega T_e\},$$

where ω_c is the crossover frequency corresponding to the gain compensation K_e of the operator performing tracking by using the display, and T_e is the effective time delay due to both reaction time and neuromuscular dynamics.

The overall performance of the model is improved by increasing the equivalent gain and reducing the equivalent time delay. The two parameters K_e and T_e describe the overall characteristics of tracking based on using this display, just as in the case of the evaluation of the mobility aid for the blind (Tachi, Mann and Rowell, 1983).

Thus, the quantity

$$\mathrm{EV}(i) = K_e(i) + \frac{1}{T_e(i)},$$

was selected for the purpose of determining and evaluating quantitatively the effectiveness of each display method, where (i) is the display method number described above.

In order to estimate the effective gain and the effective time delay, a line with a slope of $-20\,\mathrm{dB}$/decade was fitted to the amplitude of the transfer function near the crossover frequency, and by using the least squares method, the crossover frequency f_c was measured for each of the five display schemes.

The phase margin ϕ_m was measured as $\phi_m = 180 - P_c$, where P_c is the phase value at the crossover frequency. The effective gain K_e and the

Human Tracking Characteristics

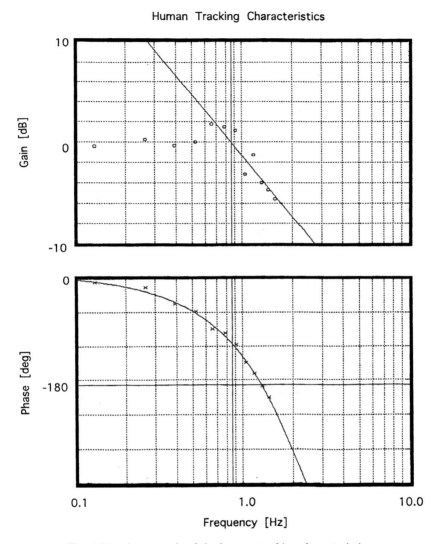

Fig. 4.25. An example of the human tracking characteristics.

effective time delay T_e were calculated from the following formulae:

$$K_e = 2\pi f_c,$$

$$T_e = \frac{1}{K_e}\left(\pi - \phi_m \cdot \frac{\pi}{180}\right).$$

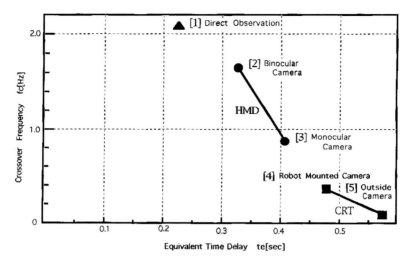

Fig. 4.26.   Comparison results.

Figure 4.26 shows the results for an operator using each of the five display schemes. These two parameters f_c and T_e are plotted in Fig. 4.26, which clearly shows the superiority of HMDs over conventional CRTs. When HMDs are used, an operator can control the directions (pitch, roll, and yaw) of the slave camera in accordance with the movements of his or her head, while the size and the perspective view of the slave manipulator and the target are the only cues in the case of CRTs. This is the main reason for the observed differences.

In the HMD group, the binocular display is associated with higher performance than the monocular display. Since no translational movement of the slave camera is allowed in this experiment, the superiority of binocular display is expected. Although we did not conduct a quantitative evaluation for the monocular display, which allows the translational motion of the slave camera following the movements of the human operator's head, our preliminary experiments suggested that the performance of monocular HMD with respect to translational movement improved with the use of motion parallax. The effects of motion parallax need further quantitative evaluation.

The differences between the CRT groups are due to the arrangement of the cameras and the slave manipulator. If we limit the arrangement to only the type described in condition 5, i.e. if the line of sight of the camera coincides with the direction of linear movement of the target, this result clearly indicates the effectiveness of the natural arrangement of the cameras and the manipulator close to the positional relations of the human eyes and the arm used for manipulation.

The experiment was conducted with five operators (all males in their thirties), and they showed the same tendency, although their absolute evaluation values were different. Figure 4.27 shows the normalized averaged evaluation value of each display scheme. The performance under direct observation was used as a standard, and its value was set to 1.

Figure 4.28 shows the comparison results, in which the root mean square error of the output from the input was used as a criterion for

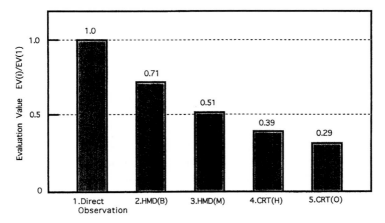

Fig. 4.27. Comparison result using the EV criterion.

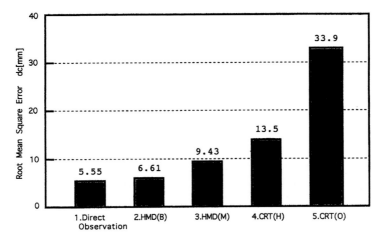

Fig. 4.28. Comparison results based on using the root mean square value.

comparison. The comparison results showed the same tendency as that shown when the EV criterion was used.

The HMD (B) display type results were superior to all other conditions except direct observation. In this regard, the method of observation was not the only difference between the direct observation and the HMD (B) observation. The use of a master manipulator as the tracking manipulator means that the direct observation is free from the disadvantageous effects of the dynamics of the slave manipulator. Future experiments are necessary in order to eliminate these differences.

4.3.3. *Summary*

An experimental telexistence system was realized which enables a human operator to experience the sensation of being present in a remote real environment where a surrogate robot is located. A telexistence master–slave system for remote manipulation experiments was designed and developed, and an evaluation experiment of the telexistence master–slave system was conducted. By comparing the telexistence master–slave system with conventional master–slave systems, the efficacy of the telexistence master–slave system was verified, and the superiority of the telexistence method was demonstrated through tracking experiments.

The comparison results revealed the clear superiority of binocular vision combined with the natural arrangement of the head and the arm, which is the most important characteristic feature of telexistence.

4.4. Construction of Virtual Haptic Space (Tachi *et al.*, 1994; Hoshino and Tachi, 1998)

A method is proposed for the construction of a virtual haptic space driven by the same environment model of the real world as that of visual space. Human limb motion is measured in real time, and the subspace of the total haptic space, which is or will be in contact with the human fingertip, is constructed by using a haptic space display device. Its end effector is a device whose shape is specially designed to approximate several shapes by changing its contact sides. Its position and orientation is controlled by a pantographic mechanism referred to as active environment display (AED). The shape of the haptic space is approximated by the environment shape approximation device (SAD), and the inertia, viscosity, and stiffness of the haptic space are generated by the use of an AED controlled with mechanical impedance.

The system can represent haptic space with edges, vertices, and surfaces by using mechanical impedance information. The system is designed to provide not only feedback regarding the force, but also shape information regarding the contact object during prolonged interaction with surfaces of various shapes and mechanical impedance. The user wears a lightweight passive 7-DOF goniometer in the shape of an exoskeleton, which measures the position and orientation of the fingertip.

4.4.1. *Construction Method for Encounter-Type Virtual Haptic Space*

Several efforts have been made for the construction of virtual haptic space. Most of the developed displays are fundamentally force/torque displays, and very few are actual shape displays. The construction of a shape display for objects with continuous surfaces (Hirota and Hirose, 1993) and the representation of shapes by using robots (McNeely, 1993) are examples of successfully developed shape displays. However, these displays are unable to represent objects with arbitrary edges and vertices. In this section, a method for representing haptic space with arbitrary edges, vertices, and surfaces by using mechanical impedance information is proposed, and experimental hardware is constructed in order to demonstrate the feasibility of the method.

We restrict our consideration to the case where the virtual haptic space can be touched at one point, i.e. at the fingertip. We also restrict ourselves to the condition that we abandon the representation of the surface texture. Then, the shape of an object in the virtual haptic space can be represented as a function of the point of contact in three dimensions (x, y, and z) in the world Cartesian coordinate system. One of the attributes of virtual haptic objects is the type of fundamental shape elements the point of contact belongs to, e.g. surface, edge or vertex, together with the normal vector of the surface at the point if it belongs to a surface, or the direction vector of the edge if it belongs to an edge.

Another attribute of the virtual object which we consider in addition to its shape is its mechanical impedance, i.e. the inertia, viscosity, and stiffness for three translational directions and three rotational directions.

Figure 4.29 shows how an object in the real environment is presented in the virtual haptic space. The movement of the upper human limb is

Shape Approximation Device (SAD)

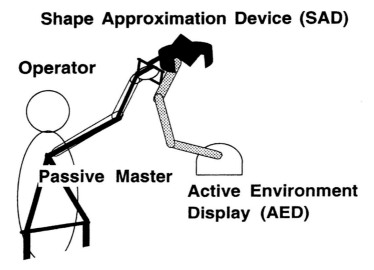

Fig. 4.29.   Conceptual diagram of the virtual haptic space construction method.

measured with a passive master arm, and the tip position and orientation of the human finger is calculated. The measured position is sent to a computer, which calculates the nearest object in the virtual haptic space. The information regarding the object, i.e. the mechanical impedance, tangential surface, and/or edge/vertex data is represented by the proposed device, which comprises a 6-DOF AED controlled by using the impedance and a SAD.

When the fingertip is in free space, no contact is made with the SAD. However, the SAD continues to display the appropriate shape information at the point nearest to the operator's fingertip. The AED follows the SAD and is controlled to locate the appropriate position on the basis of the measurement of the fingertip position/posture and the model of the virtual haptic space. When the fingertip makes contact with the point on the virtual object, the human fingertip makes contact with the SAD with the appropriate mechanical impedance provided by the AED. When the operator moves his or her fingertip, he or she feels the shape of the contact area, whether it is an edge, a vertex, or a part of the surface represented by the appropriate part of the SAD. If it is an edge, he or she can find which direction of the line in the virtual haptic space. When it is a surface, the surface orientation is represented.

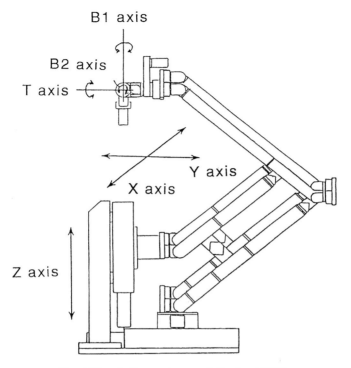

Fig. 4.30. Active environment display (AED).

This method of construction of haptic space is sometimes referred to as an encounter-type haptic display method (Yokokohji *et al.*, 1996).

4.4.1.1. *Active Environment Display*

Figure 4.30 shows the AED. It has a pantographic link mechanism with a link ratio of 1:3. Displacements in the x and y directions are magnified three times, and the displacement in the z direction is magnified four times. Auxiliary links are used in order to represent the orientation of the SAD at the end point of the mechanism independently of the position of the endpoint. This reduces the burden of the calculations necessary for the control.

The range of the display is ±300 mm for each of the x, y, and z directions. Regarding the rotational ranges, they are ±180° for yaw, ±90° for roll, and +90 to −45° for pitch.

4.4.1.2. *Shape Approximation Device*

Figure 4.31 shows the SAD designed in the present study, and Figs. 4.32 and 4.33 show how surfaces and edges are approximated by using the device. Continuous surfaces are approximated by the tangential plane at the representation point. As the contact point moves, the tangential plane

Fig. 4.31.   Shape approximation device (SAD).

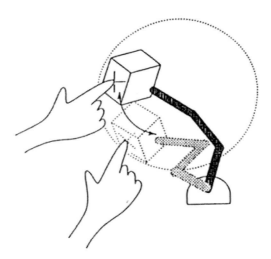

Fig. 4.32.   Surface representation.

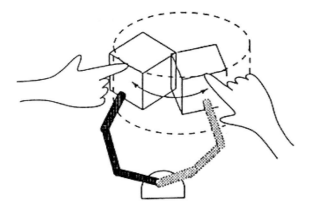

Fig. 4.33.   Edge representation.

follows the point, changing its orientation in accordance with predetermined information.

In order to approximate shapes in the environment, the approximation device must represent both convex and concave edges. Therefore, the device has both convex and concave edges, and they are used for the representation of edged surfaces (Fig. 4.31). By controlling the orientation and moving the device along a direction normal to the contact point surface normal, it is possible to arrange any edge at a predetermined position with a predetermined orientation, as well as to turn the device around the presented edge in order to construct predetermined contiguous surfaces.

4.4.2.   *Test Hardware*

Figure 4.34 shows the constructed experimental hardware system, and Fig. 4.35 indicates the block diagram of the system. Each object in the virtual space is represented in two ways. One is a geometrical representation using polygons as used in virtual visual space, and the other is represented using a sphere, a cylinder, a cone, a generalized cone, a cube, a parallelepiped, and a combination of these shapes. The object is represented by using the local coordinate fixed to the object in both cases. The position and the orientation of the origin of the local coordinate is assigned relative to the world coordinate, and each point on the virtual model is calculated with reference to the world coordinate. Visual information and haptic information are driven by the same world model, and the visual rendering

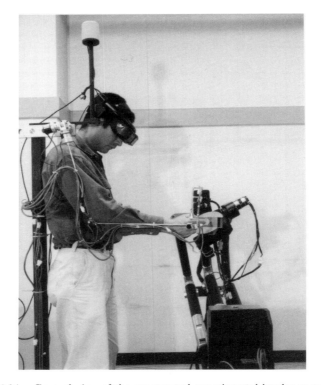

Fig. 4.34.   General view of the constructed experimental hardware system.

and the haptic rendering are conducted in parallel. The nearest part of the virtual object to the finger is estimated by a method similar to the z-buffer method.

Figure 4.36 shows the passive master device with 7-DOFs used for measuring the position and the orientation of a human user's fingertip. The human finger is covered with a brace with a ball point in order to ensure the point contact with the SAD. The brace also acts as an insulator for the human sensation of touch from the movements of the SAD upon contact.

Figure 4.37 shows an example of surface and edge representation by using AED with SAD. In this model, an object in the virtual environment is divided into quadrangles referred to as patches, and the shape of each patch is described using Bezier surfaces, which are suitable for describing complicated shapes.

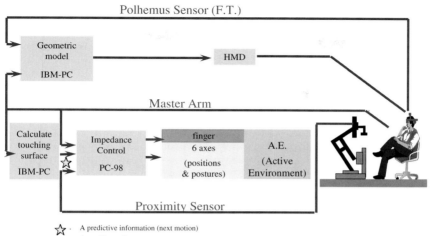

Fig. 4.35.   Block diagram of the virtual visual and haptic space representation system.

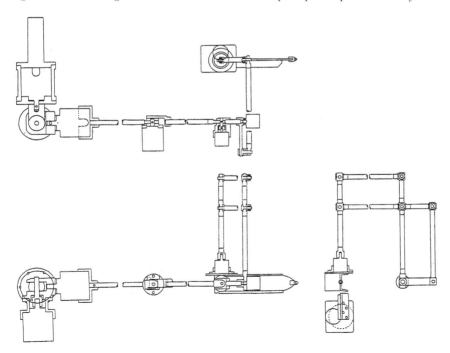

Fig. 4.36.   Passive master system.

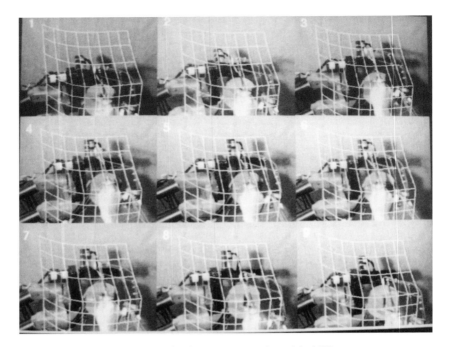

Fig. 4.37.   Surface representation with AED.

4.4.3. *Summary*

A method for constructing virtual haptic space was proposed, and experimental hardware was prepared on the basis of the proposed method. It was shown that an object with vertices, edges, and surfaces can be correctly represented. Both concave and convex edges were successfully represented by using the test hardware.

A 6-DOF manipulator, which we refer to as "AED," moves in an anticipatory manner and, if necessary, counters the user's movements in order to represent contact with virtual surfaces, edges, and vertices. The robot manipulator is a pantograph with a range of motion of $60 \times 60 \times 60$ cm for translation and 360, 180, and 135° for yaw, roll, and pitch, respectively.

The wrist of the manipulator carries a device with a complex surface geometry including convex and concave edges and flat surfaces. It is thus possible to simulate contact with continuous surfaces and edges by moving the SAD. Contiguous surfaces are displayed by reorienting the corresponding facet of SAD. The user wears a head-mounted display, which immerses him or her into the virtual environment.

Thus, the user can feel exactly what he or she sees at the observed position by using the haptic shape. Experiments were conducted where several shape objects were represented with several kinds of mechanical impedance. It was shown that an object with vertices, edges, and surface can be represented, where both concave and convex edges are represented successfully using the test hardware.

Chapter 5

Retroreflective Projection Technology (RPT)

The enhancement of our ability to perceive the world around us has long been a dream of mankind, and it is finally being attained through augmented reality (AR). This multi-disciplinary field endeavors to seamlessly integrate digital information into real environments. The author and his team proposed and developed retroreflective projection technology (RPT), which is an elegant and cost-effective approach to turning any real object into a strikingly lifelike visual display. Retroreflective materials reflect light back in the direction of the source, which can be employed to produce new versatile optical systems supporting stereoscopic vision and accurate occlusion. Since the projection surface can be of any shape, one practical application referred to as "Optical Camouflage" realistically projects a background image onto an opaque object, such as a vehicle or an overcoat, in order to make it appear invisible; a concept previously confined to science fiction.

This technology can be used in various applications. For example, if used in a cockpit to make the floor of the airplane transparent during landing, the pilot can perform a safe landing with the runway in full view. Similarly, RPT can be used to eliminate the notorious "blind spot" for trucks and automobiles. In addition, it can provide an outside view for rooms with no windows, in which the wall would appear to be invisible. Pre-captured X-ray and/or MRI data can be superimposed onto the body of a patient, providing the surgeon with essential information for open surgery even under minimally invasive surgical environments. By superimposing ultrasonic data, a real-time representation of the inner parts of the body is also possible through RPT.

This technique has evolved from studies aimed at realizing the seamless integration of digital information into real environments in order to improve our safety, boost our productivity, and enhance our ability to perceive the world around us.

5.1. Principle of Retroreflective Projection Technology

Two classic types of visual displays for virtual reality are the Head-Mounted Display (HMD) (Sutherland, 1968; Fisher *et al.*, 1986) and Immersive Projection Technology (IPT) (Cruz-Neira *et al.*, 1993), which, although quite useful, are not without their shortcomings, as shown in Figs. 5.1(C) and (D), respectively. The former suffers from the tradeoff of high resolution and wide field of view, and the latter faces problems concerning the user's body casting shadows in the virtual environment as well as difficulties with the interaction between the user's real body and the virtual interface. In addition, both displays encounter problems concerning occlusion when in use under the augmented reality condition, i.e. when virtual objects and real objects are mixed.

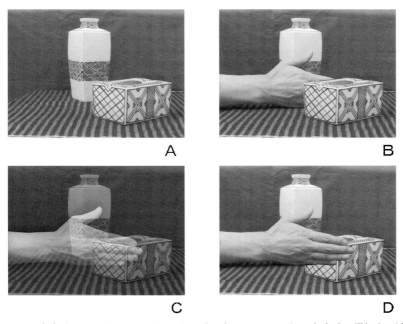

Fig. 5.1. (A) A virtual vase and a virtual ashtray on a virtual desk. (B) An ideal occlusion, where a real hand is placed between two virtual objects. (C) Unfavorable results when an optical see-through HMD is used. (D) Unfavorable results in the case of IPT (Immersive Projection Technology) implementations, such as CAVE.

Figure 5.1(A) shows a virtual vase and a virtual ashtray on a virtual desk. When a real hand is placed between two virtual objects, in the ideal case, occlusion should be realized as shown in Fig. 5.1(B), i.e. the real hand occludes the virtual vase while being occluded by the virtual ashtray at the same time. However, a real hand cannot occlude the virtual vase, nor can it be occluded by the virtual ashtray when an optical see-through HMD is used to display the virtual objects, and both the hand and the ashtray appear to be transparent. This is simply due to the fact that the position of the physical display of an HMD is always just in front of the eyes of the user.

Conversely, the virtual ashtray cannot occlude a real hand in the case of IPT implementations, such as CAVE (CAVE Automatic Virtual Environment), as shown in Fig. 5.1(D). This is because the displayed position of the virtual objects is always on the screen surface, which is one to two meters away from the human user in the case of IPT displays.

A new type of visual display is currently being developed under the names "media X'tal" (pronounced crystal) (Kawakami *et al.*, 1998) and "X'tal vision" (Inami *et al.*, 1999, 2000), which uses retroreflective materials as its projection surface. This type of display technology is comprehensively referred to as RPT (Tachi, 1999b, 2003b).

Under the RPT configuration, a projector is placed at the axial symmetric position of the user's eyes, with a half mirror used as a reference, and a pinhole is placed in front of the projector in order to ensure adequate depth of field, as shown in Fig. 5.2. This condition is referred to as the

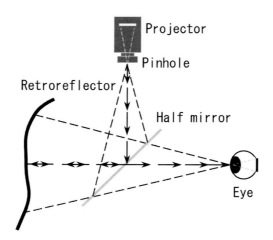

Fig. 5.2.   Principle of RPT system.

"conjugate optical condition," which is the principle of RPT. Images are projected onto a screen which is constructed, painted, or covered with a retroreflective material.

While conventional screens used in IPT scatter projected light equally in all directions, similarly to a Lambertian surface, retroreflective surfaces reflect the projected light only in the direction of projection (Fig. 5.3).

Figure 5.4 shows how a retroreflective surface behaves. It is covered with microscopic beads about 50 micrometers in diameter, which reflect the incident light back to the direction of incidence. This effect can also be achieved by using a microstructure of prism-shaped retroreflectors densely distributed onto the surface.

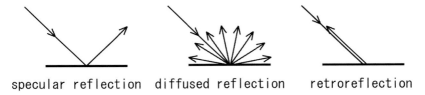

specular reflection   diffused reflection   retroreflection

Fig. 5.3.   Three typical reflection patterns.

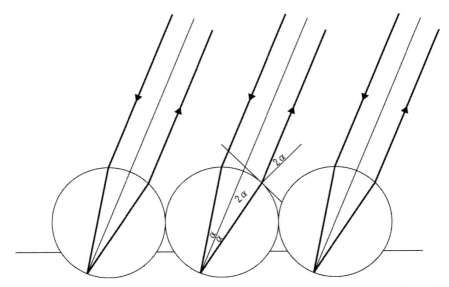

Fig. 5.4.   Retroreflective surface densely covered with microscopic beads about 50 micrometers in diameter. Ideally, the refractive index should be 2.

The retroreflector screen, together with the pinhole, ensures that the user always sees images with accurate occlusion relations. Arbitrary screen shapes can be used in the construction of RPT systems due to the characteristics of retroreflectors and the presence of a pinhole in the conjugate optical system.

By using these characteristics of RPT systems, binocular stereovision becomes possible by using only one screen with an arbitrary shape. Figure 5.5 shows how stereovision can be realized by using RPT. In the figure, the Display Unit is an arbitrarily shaped object covered or painted with a retroreflective material. The light projected by the right projector is retroreflected from the surface of the display unit and is observed by the right eye, while the light projected by the left projector is retroreflected by the same display surface and can be observed only by the left eye.

By using the same display surface, the right eye observes the image projected by the right projector and the left eye observes the image projected by the left projector. Thus, by generating CG images with the appropriate disparity, the human observer perceives a stereoscopic view of the projected objects at the position of the display unit. By using measurements from the position sensor located on the display unit, it is possible to display a 3D image of the object, which changes its appearance in accordance with the position and the orientation indicated by the motion of the display. This allows the user to experience the sensation of handling a real object.

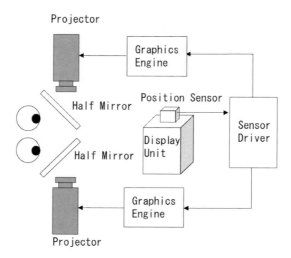

Fig. 5.5.   Principle of a stereo display based on RPT.

The same technology can be used by multiple users to observe the same screen. Multi-projection of images on the same screen is possible. Each user projects his or her image onto the screen by using RPT while retaining the ability to see his or her image independently. This characteristic feature can be used, for instance, for projecting multiple images onto the same robot, as in the case of mutual telexistence (Tachi, 2003b; Tachi *et al.*, 2008). This application will be explained in detail in Chap. 6.

5.2. RPT-Based Head Mounted Projector

A type of projector referred to as a "RPT-based Head Mounted Projector (HMP)" can be mounted on the head of a user. Figure 5.6 illustrates the implementation of RPT on an HMP, and Fig. 5.7 shows a general view of a prototype HMP. However, although an HMP based on RPT makes it possible to project stereoscopic color images onto a screen free-form surface with high contrast and brightness, a large half mirror covers most of the wearer's face.

In this regard, X'tal Visor was developed as a full open-type RPT-based HMP (Sonoda *et al.*, 2005). The X'tal Visor replaces the half mirror with an all-reflective micro mirror. The mirror is placed near the focal point of the projection lens, and therefore the minimal required size of the mirror becomes negligibly small. Since the mirror is placed quite near to the user's eye, the conjugate optical condition necessary for RPT is attained without using a half mirror. Figure 5.8 illustrates the principle

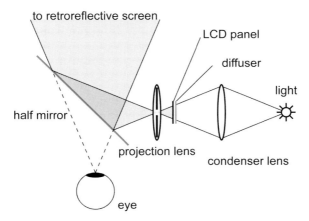

Fig. 5.6.   Principle of an original RPT-based HMP (Head Mounted Projector).

Fig. 5.7.   General appearance of a Head Mounted Projector (HMP).

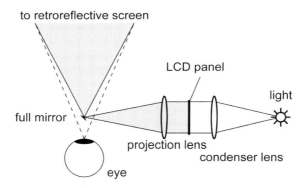

Fig. 5.8.   Principle of a full open-type HMP based on RPT.

of an HMP using a micro mirror. Since the size of the mirror is smaller than the size of the user's pupil, it does not hinder his or her field of view. X'tal Visor can display images without covering the wearer's face, and therefore enables natural open face-to-face communication. In addition, a spherical mirror can be used to attain a wider field of projection. Since spherical mirrors distort the projected image, the original image on the LCD screen must be pre-modified in order to attain an undistorted image on the screen. Figure 5.9 shows a general view of an implementation of X'tal Visor.

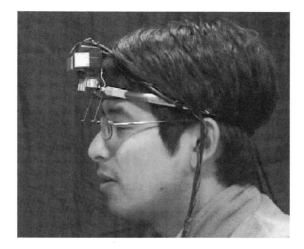

Fig. 5.9.   General view of X'tal Visor.

5.3. RPT Applications

Figure 5.10 shows an example of an image projected onto a person wearing a shirt covered with a retroreflective sheet. As apparent in the figure, the projected image looks like a real skeleton, where the image is partly occluded by the fingers of the person wearing the retroreflective material.

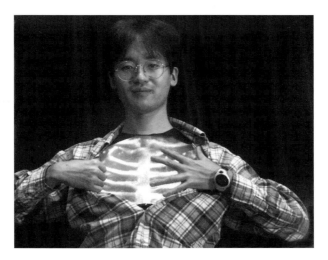

Fig. 5.10.   Image projected onto a retroreflective screen.

This is an example of the use of RPT for augmented reality. Pre-captured X-ray and/or MRI data can be superimposed onto a human patient to provide the surgeon with essential information for open surgery even under minimally invasive surgical environments. By superimposing ultrasonic data, the real-time representation of the inner parts of the body is also possible through RPT.

Figure 5.11 shows an example of projecting a virtual cylinder onto a Shape Approximation Device (SAD), which is a haptic device (explained in Sec. 4.4) that allows the user to experience a realistic tactile feeling when touching virtual objects. The use of SAD as a retroreflective screen allows the user to feel as though observing the object through an HMP. Figure 5.11(A) illustrates the principle of SAD, Fig. 5.11(B) shows the displayed image, Fig. 5.11(C) is an actual SAD device, and Fig. 5.11(D) indicates the image projected onto SAD, in which tactile information corresponding to the shape of the object can be perceived correctly.

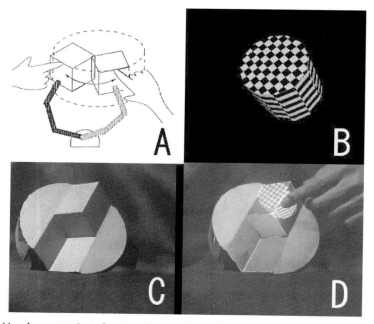

Fig. 5.11.   Image projected onto a Surface Approximation Device (SAD). (A) Principle of AED (Active Environment Display) based on SAD. (B) Image. (C) Actual SAD. (D) Image projected onto SAD, in which tactile information corresponding to the shape of the object can be perceived correctly.

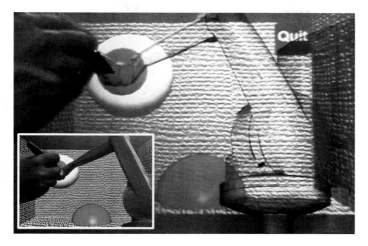

Fig. 5.12.   An example of the application of RPT in medicine.

Fig. 5.13.   Optical camouflage using RPT.

Figure 5.12 shows an example application of RPT for "optical camou-flage". In Fig. 5.12, a pre-captured background image is projected in such a way that a retroreflective object held by the user appears to be transparent.

Figure 5.13 shows how "optical camouflage" can be achieved by using real-time video information. Figure 5.14 shows how RPT is utilized for the

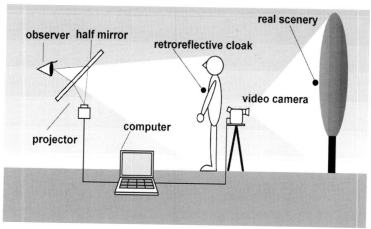

Fig. 5.14. Schematic diagram of the RPT system used for optical camouflage in Fig. 5.13.

realization of the situation of Fig. 5.13. The coat is made of a retroreflective material so that light is reflected back to the direction of incidence. Microscopic beads on the surface of the coat provide the functionality of retroreflection, as was explained previously.

An half mirror allows observers to see virtually from the position of the projector. An HMP projects an image of the background scenery captured by the video camera behind the camouflaged person. A computer uses image-based rendering techniques to calculate the appropriate perspective and transform the captured image into the image to be projected onto the subject. Since the cloak the subject is wearing is made of a retroreflective material, which reflects the incident light in the direction of incidence, an observer looking through a half mirror sees a very bright image of the scenery behind the person wearing the camouflage material, as if he or she is virtually transparent.

Another example is a "transparent cockpit," as is shown in the case of a passenger car in Fig. 5.15. For the purpose of safety and increased operability, it is important to obtain a wide field of view while operating a vehicle. However, the space for windows is limited. Therefore, the concept of a "transparent cockpit," in which an image of the objects located in a blind spot is displayed on the inner wall of the vehicle by using RPT is quite promising. In this system, the internal components of the vehicle, such as the doors and the floor, are virtually transparent, the blind spot is

Fig. 5.15. Transparent cockpit of a passenger car.

eliminated, and objects which would otherwise be occluded by structural elements are observed as if through a window (Yoshida *et al.*, 2008).

This technology can be used, for instance, in helicopters or airplanes, whose floor can be made virtually transparent by using RPT, which can aid the pilot in landing more easily.

Thus, RPT can provide a way to transform any physical object into a virtual object simply by covering its surface with a retroreflective material.

Chapter 6

Mutual Telexistence Using RPT

Telexistence technology provides a highly realistic sensation of presence in a remote place without requiring any actual travel. The concept was first proposed by the author in 1980, and its feasibility has been demonstrated through the construction of alter-ego robot systems such as Telesar (TELExistence Surrogate Anthropomorphic Robot), which was developed under the national large-scale project entitled "Robots in Hazardous Environments," as well as the HRP supercockpit biped robot system developed under the "Humanoid Robotics Project." Telesar II, a mutual telexistence system, was subsequently developed and can generate the effect of existing in a remote place in local space by the combined use of an alter-ego robot and retroreflective projection technology (RPT). Thus, the feasibility of mutual telexistence has been demonstrated. In this chapter, the concept of mutual telexistence using RPT is described, the design and the construction of the Telesar II system are explained, and the feasibility and the efficacy of the mutual telexistence system are experimentally demonstrated using Telesar II.

6.1. Mutual Telexistence

It was back in 1980 when the author initially proposed the concept of telexistence, and the function of telexistence providing a sensation of presence in remote and/or virtual environments has since been realized. We can now work and complete physical actions with the realistic feeling that we are present in several places at once.

By using a telexistence system, a user can control a robot by simply moving his or her body naturally, without the need for verbal commands. The robot conforms to the user's motion, and through sensors on board the robot, the human can see, hear and feel as if experiencing the remote environment directly. In this manner, people can virtually exist in a remote environment.

For observers in the remote environment, however, the situation is quite different: they see only the robot moving and speaking. Although they can hear the voice and witness the behavior of the human operator through the robot, it does not actually look like him or her. This means that the telexistence is not yet mutual.

Simply placing a TV display on board the robot to show the user's face is not very satisfying since the face does not change with the observation angle, and thus appears largely comical and far removed from reality. By using RPT in conjunction with a head-mounted projector (HMP), the mutual telexistence problem can be solved, as shown in Fig. 6.1.

Suppose user A uses his telexistence robot A' at a remote site where another user B is present. User B in turn uses another telexistence robot B', which exists at the site where the user A works. 3D images of the remote scenery are captured by cameras on board both robots A' and B', and are sent to the HMP's of users A and B, respectively, providing both users with a sensation of presence. Both telexistence robots A' and B' are seen as if they were their respective human users by projecting the real image of the users onto their respective robots. However, this situation is somewhat confusing because two real environments [A] and [B] are involved.

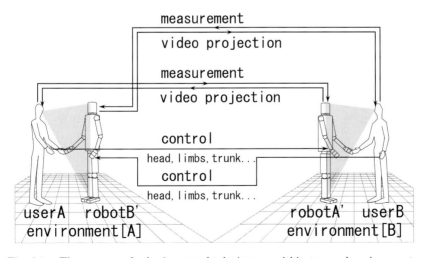

Fig. 6.1.   The concept of robotic mutual telexistence within two real environments.

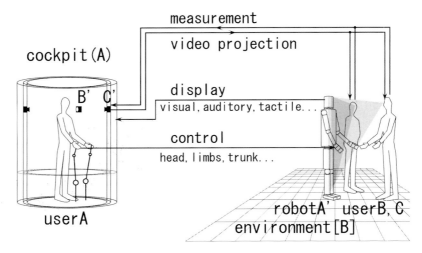

Fig. 6.2. The concept of robotic mutual telexistence with one real environment.

As shown in Fig. 6.2, this can be simplified further by restricting the real environment to just one environment, i.e. environment [B]. In this situation, only user A in cockpit (A) can telexist using his surrogate robot A′, while users B and C remain in their real environment [B].

Cockpit (A) has a capability of displaying 3D images of environment [B] in 360° around user A, who can enjoy the scenery of real environment [B] without using any eyewear, i.e. autostereoscopically. This can be realized using technology called TWISTER (Telexistence Wide-angle Immersive STEReoscope), which will be described in Chap. 7. Using control devices installed in the cockpit, user A can control his or her robot with a sensation of presence in real environment [B]. These control devices are haptic devices like the ones described earlier in Sec. 4.4 and later in Sec. 6.3, and a device using Gravity Grabber (Minamizawa *et al.*, 2007). Users B and C can, respectively, project images taken by cameras B′ and C′ onto robot A′, thereby enabling them to see user A as if he or she exists inside the robot.

The cameras are virtually controlled in order that users B and C, respectively, see from a direction relative to user A. Position measurements of users B and C are made by the robot, and the appropriate directional images of user A are transmitted to users B and C through the robot, respectively. Thus, mutual telexistence is realized.

6.2. Experimental Hardware System

In order to realize mutual telexistence, the author and his team have been pursuing the use of projection technology utilizing a retroreflective material as a surface. The idea was first proposed by Tachi (1999b).

Figure 6.3 presents an example of how mutual telexistence can be achieved through the use of RPT. Suppose user A uses his telexistence robot A′ at a remote site where user B is present. User B in turn uses another telexistence robot B′, which exists in the site where user A works. 3D images of the remote scenery are captured by cameras on board both robots A′ and B′, and are sent to the respective HMPs of the users, thereby providing them both with a sensation of presence. Both telexistence robots A′ and B′ are seen as if they were their respective human users by projecting the real image of the users onto their respective robots.

Figure 6.4(A) shows a miniature of the HONDA humanoid robot, whereas Fig. 6.4(B) shows the robot painted with retroreflective material. Figures 6.4(C) and 6.4(D) show how they appear to a user wearing an HMP. The telexisting robot looks just like the human operator of the robot,

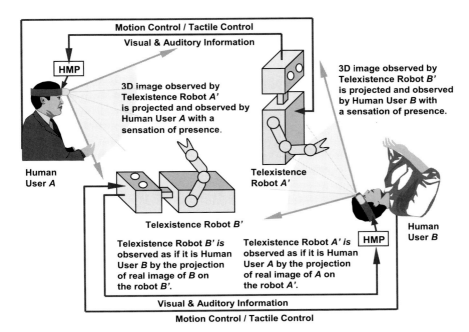

Fig. 6.3.   The concept of robotic mutual telexistence.

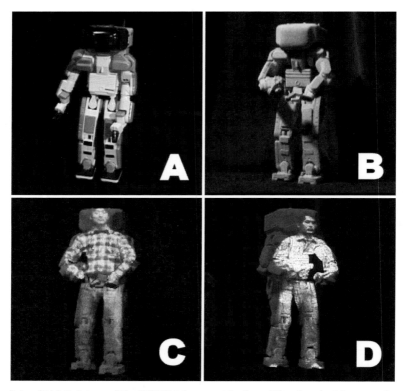

Fig. 6.4.   (A) Miniature of the HONDA humanoid robot, (B) painted with retroreflective material, and (C) and (D) with examples of a human projected onto it.

and telexistence can be naturally performed (Tachi, 1999b). However, this preliminary experiment was conducted off-line.

6.2.1. *Preliminary Mutual Telexistence*
Hardware System (Tachi et al., 2004)

In order to verify the feasibility of the proposed method on-line, an experimental hardware system was constructed (Fig. 6.5). In the figure, user A tries to telexist in remote environment [B] from local cockpit (A) using robot A′. User A is in local cockpit (A) where his head motion is measured by ADL1 (Shooting Star Technology, Inc.), a mechanical goniometer with six degrees of freedom (DOFs). He observes remote environment [B] through a back projection display in the cockpit, while his own figure is captured by

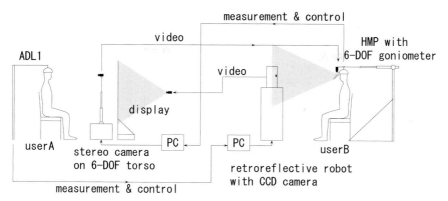

Fig. 6.5. Schematic diagram of the robotic mutual telexistence system experimentally constructed.

a stereo camera mounted on a specially designed and constructed 6-DOF torso servomechanism.

In remote environment [B], a robot equipped with PA-10 (Mitsubishi Heavy Industry Co.) as a head motion mechanism is covered with retrore-flective material. Images captured by a camera inside the retroreflective robot's head are sent to the rear projection display in the local cockpit. User B sees the retroreflective robot using an HMP. His head movement is measured by a specially designed and constructed 6-DOF counterbalanced position/orientation measurement system. User B's head motion is transmitted to the local cockpit, where a torso stereo camera is controlled in accordance with the tracked motion of user B. Figures 6.6 and 6.7 show the torso stereo camera mechanism. It has six DOFs and is designed to track seated human motion at frequencies up to 1.3 Hz. Two parallel cameras are placed 65 mm apart from each other, each with a horizontal field of view of 45°.

Figure 6.8 shows the mechanism of the 6-DOF goniometer for HRP, and Fig. 6.9 shows its general appearance. The weight of the HRP (1.65 kgf) is fully counterbalanced by a weight and spring, while six degrees of head motion (up/down, left/right, back/forth, pitch, roll, and yaw) are fully unrestricted. For the positioning, spherical coordinates are used with translational motion of 980–1580 mm, base pitch of −15–15°, and base yaw of −180–180°, while the orientation is realized using a three-axis pivot with pitch of −60–60°, roll of −90–90°, and yaw of −30–30°.

Figure 6.10 shows the general appearance of the constructed HMP, which consists of two 0.7-inch full-color LCD projectors with a resolution

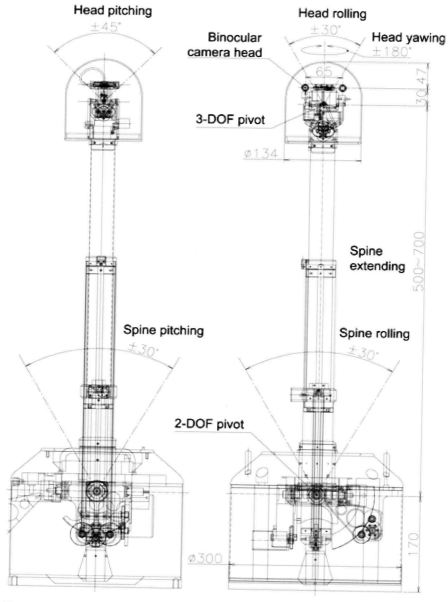

Fig. 6.6.   Design of stereo camera system mounted on a torso mechanism with six DOFs.

Fig. 6.7.   General appearance of the torso stereo camera mechanism with six DOFs.

of 832×624, two pinholes, and an acrylic half mirror. The horizontal field of view of the projector is $60°$.

Figure 6.11 shows the dimensions and the mechanism of the constructed retroreflective robot. The torso of the robot is fixed and does not move; however, its head can move up and down, left and right, back and forth, and rotate, pitch, roll and yaw through the use of a robot manipulator PA10. The robot is completely covered with retroreflective material, including the bellows connecting its head and torso.

A video camera is mounted on top of the mechanism 32.5 mm shifted from the center, where a hole of 7 mm diameter is open on the surface of the head. The captured image is sent to user A. The motion of the retroreflective robot is controlled to follow user A's motion.

An example of the experimental results is shown in Fig. 6.12. User A is in local cockpit (A) and his motion is measured by ADL1 (Figs. 6.12(1a) and 6.12(1b)), and he moves to the left in Fig. 6.12(1b). In Figs. 6.12(2a) and 6.12(2b) user B and telexistence robot A′ are facing each other in environment [B]. The robot moves to the left according to the motion of user A. Figures 6.12(3a) and 6.12(3b) show the image of user A projected onto robot A′.

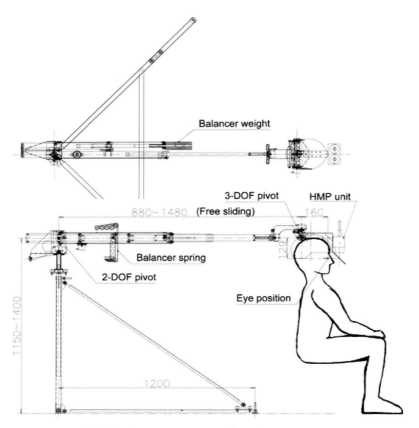

Fig. 6.8.   Design of the head-mounted projector (HMP) suspended by a counterbalanced goniometer with six DOFs.

The image is retained on the robot's head when it is moved to the left. The black dot on the surface of the robot's head indicates the location of the camera. The head is controlled such that the point always coincides with the location of the left eye of user A.

6.2.2. *Summary*

Mutual telexistence is one of the most important technologies for the realization of networked telexistence since users "telexisted" via a robot must know with whom they are working over the network. The proposed method using RPT, especially that utilizing an HMP and a robot with

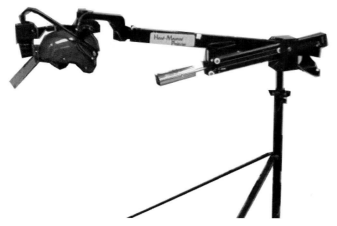

Fig. 6.9. General appearance of the HMP suspended by the counterbalanced goniometer with six DOFs.

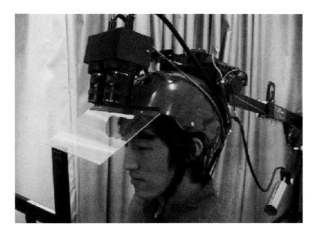

Fig. 6.10. General appearance of the HMP used in this configuration.

retroreflective covering, was proven in the preliminary experiments to be a promising approach toward the realization of mutual telexistence.

6.3. Mutual Telexistence Master–Slave System for Communication (Tachi *et al.*, 2008)

A new prototype of a mutual telexistence master–slave system for communication has been designed and developed. This system is based on RPT

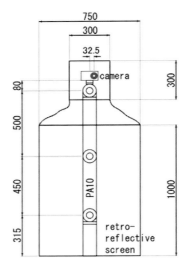

Fig. 6.11.   Configuration of the retroreflective robot.

and, as shown in Fig. 6.13, is composed of three subsystems: a Telesar II slave robot, a master cockpit, and a viewer system.

The robot constructed for this communication system is referred to as "Telesar II." In order to use this system for telecommunication, we focused our design on reproducing realistic human-like movements. Telesar II has two human-sized arms and hands, a torso, and a head. Its neck mechanism has two DOFs, which can rotate around pitch and roll axes. Two CCD cameras are located in its head for stereoscopic vision. For the benefit of an operator, it also has four pairs of stereo cameras mounted on top of the head for a 3D surround display. A microphone array and a speaker are also employed for auditory sensation and verbal communication, respectively. Each arm has seven DOFs, and each hand has five fingers with a total of eight DOFs.

To control the slave robot, a master cockpit was developed. The cockpit consists of two master arms, two master hands, a multi-stereo display system, speakers and a microphone, and cameras for capturing the images of the operator in real time. In order that the operator can move smoothly, each master arm has 6-DOF structures to free the operator's elbow from constraints. To control the redundant seven DOFs of the anthropomorphic slave arm, a small orientation sensor is placed on the operator's elbow. In this manner, each master arm can measure 7-DOF motions for each

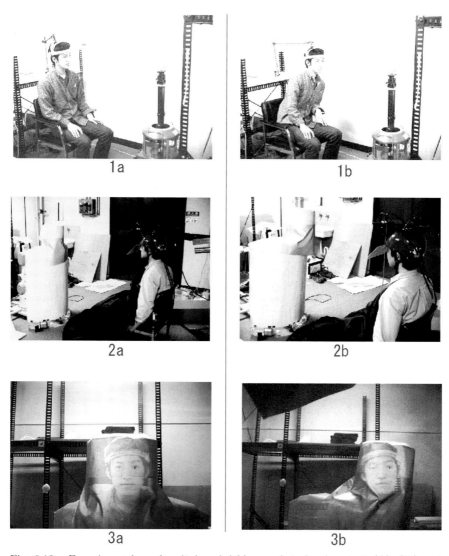

Fig. 6.12. Experimental results: (1a) and (1b) user A in local cockpit (A); (2a) and (2b) user B and telexistence robot A′ in environment [B]; and (3a) and (3b) image of user A projected onto robot A′. The black dot indicates the location of the camera.

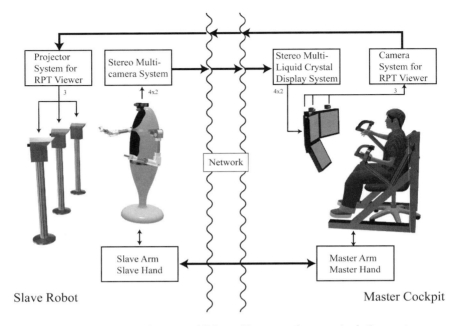

Fig. 6.13.   Schematic diagram of Telesar II master–slave manipulation system.

corresponding slave arm, while the force is fed back from each slave arm to each corresponding master arm with six DOFs.

The master arm is lightweight, and its impedance is controlled such that the operator feels as if he or she is inside the slave robot. It is important that the master should transmit the exact amount of force to the operator and that the slave robot should maintain safe contact with humans in a remote environment. The impedance-control-type master–slave system can achieve this (Tachi and Sakaki, 1992). Moreover, safe compliant contact can be maintained with humans because the slave is subject to impedance control. The motion of the robot's head is synchronized with the motion of the operator's head, which is measured by using a head tracker in the master cockpit.

The operator can easily control the hands of Telesar II since his hand motions are measured by the master cockpit manipulators and controlled by master–slave methods. In the case of an autonomous robot system, a precise computation of the motions is required in order to prevent collisions of the arms, hands, or torso of the robot. In the case of the telexistence system, however, collision detection is not necessary; the operator calculates

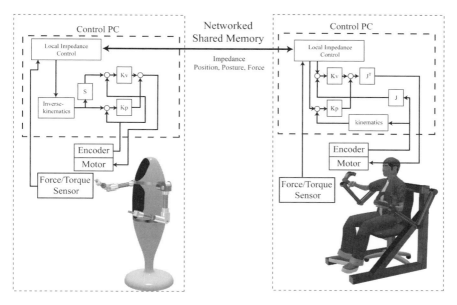

Fig. 6.14. Schematic diagram of impedance-controlled teleoperation.

it subconsciously. This is a remarkable feature of the telexistence system. Despite this, we calculated the collision limit, and collision can be prevented even in the case of operator error with a fail-safe mechanism or safety intelligence system. Figure 6.14 presents a schematic of the impedance-control-type master–slave teleoperation system used in this study.

The most distinctive feature of the Telesar II system is the use of an RPT viewer system. Both the motion and the visual image of the operator are important factors for providing a sensation of presence at the site where the robot is working. In order to view the image of the operator on the slave robot such that it looks as if the operator is inside the robot, the robot is covered with retroreflective material and the image of the operator captured by a camera in the master cockpit is projected onto Telesar II, enabling an observer to view the robot as if it were the operator.

6.3.1. *Telexistence Surrogate Anthropomorphic Robot II*

6.3.1.1. *Slave Robot Arm*

Telesar II has two 7-DOF arms, as shown in Fig. 6.15. Each arm is designed such that its weight is minimum in order that it can move rapidly and is

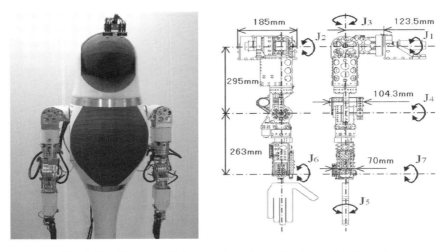

Fig. 6.15.   Slave arm (left: overview; right: structure of right arm).

safe for human use. By uniting the housing parts of a harmonic drive gear
system with other parts, such as the rotational axes of joints, it has been
ensured that the entire mechanism of the arm is lightweight, at 7.3 kgf,
with a payload of 0.5 kgf. The maximum velocity of the arm is 1.2 ms. The
slave arm supports sufficient payload and speed for mutual telexistence by
allowing the use of gestures, and since it is considerably lighter than other
available arms (Tadakuma *et al.*, 2005), the potential danger of injury due
to malfunction is also greatly reduced. Additionally, the slave arm has a
larger range of mobility in the joints above the elbow than do other arms.
Three shoulder joints J1, J2, and J3 have a mobility range of $-180°$ to
$180°$, $0°$ to $180°$, and $-180°$ to $180°$, respectively. The reduction ratio of
the harmonic drive of each joint of the slave arm is set to 50 in order to
maintain back-drivability. The maximum force at the end of the slave arm
is 164 N.

The motor driver controls the DC motors in the joints by means of
a torque control mode in accordance with commands received from a DA
board. The angular velocity and posture of each joint are measured using an
encoder attached to the motor. The neutral point of each joint is defined by
photo-interrupters, whose signals are read by the AD board. The control
system of the slave arm is connected to the master arm system through
shared memory. The control system for the master arm is the same as that
for the slave arm.

The distribution of the joints of the arm replicates the structure of the human arm in order to facilitate operation by telexistence using kinesthetic sensation. This human-mimicked structure is also useful for interaction with people since the operator can fully appreciate the sensation of congruity.

6.3.1.2. *Slave Robot Hand*

Each slave hand has five fingers. Its thumb has three DOFs, the remaining four fingers have one DOF each and the hand itself has abduction DOF, making a total of eight DOFs. The hand weighs 0.5 kgf. The size of the hand is similar to that of a typical human, with length 185 mm, width 100 mm, and thickness 35 mm. All parts such as motors, gears, and encoders are packed inside the hand mechanism. The hand is connected by a single cable to the control system, i.e. a servoamplifier and control computers.

Figure 6.16 shows the general appearance of the slave hand used in our study.

Natural teleinteraction with remote objects is realized once the operator acquires haptic information correctly through the slave hand. It is therefore desirable that the slave hand has a human-like structure with appropriate haptic sensors. To this end, a finger-shaped haptic sensor using GelForce technology was developed. GelForce is a haptic sensor, which measures the distribution of both the magnitude and the direction of force (Kamiyama *et al.*, 2004). This sensor comprises a transparent elastic body, two layers of blue and red markers, and a CCD camera (Fig. 6.17). The distributed force applied to the finger surface causes the deformation of the elastic body,

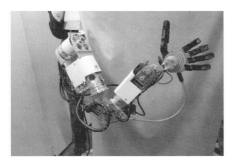

Fig. 6.16.   Slave hand (left: right hand with right arm; left: right-hand palm).

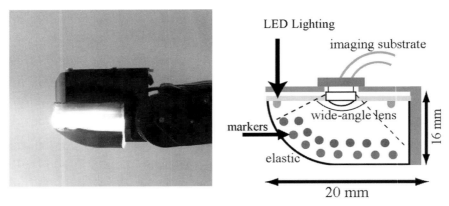

Fig. 6.17.   Finger-shaped GelForce mounted on the fingertip of the dexterous slave hand
(left: general view; right: finger-shaped GelForce sensing mechanism).

which changes marker patterns. The pattern of deformation is measured
by a camera, and a computer estimates the distributed vector forces from
the measured pattern by using an inverse algorithm. The finger-shaped
GelForce sensor achieves a space resolution of 4.5 mm and a time resolution
of 14 ms. These performances can be further improved by enhancing the
performance of the camera frame rate.

The most remarkable feature of this method is that it can measure both
distribution and direction of the force, enabling the shearing forces to be
measured (Fig. 6.18).

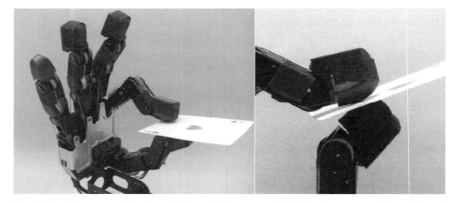

Fig. 6.18.   Dexterous slave hand (left: pinching a card with the fingertips; right: enlarged
view).

6.3.2. *Telexistence Cockpit*

6.3.2.1. *Master Arm*

The slave arm of Telesar II, alongside a conventional anthropomorphic slave arm, both of which have the same structure as a human arm, is constructed with 7-DOF mechanisms. The master arm used as the teleoperation system is also usually constructed with a 7-DOF structure. However, it is normally difficult for the operator to achieve free motion if the seven DOFs are active since it tends to restrain the operator's elbow mechanically. When the force to be applied to the operator's hand is considered, it is along a maximum of six axes, i.e. six DOFs, although the motion of the human arm has seven DOFs. Thus, the master arm has been designed such that it effectively performs the function of force presentation in these six axes; in other words, the master arm has been constructed as a 6-DOF mechanism.

While the force feedback mechanism is sufficient for six DOFs, it is necessary to have seven DOFs for the measurement of human arm motion. The master arm has a cantilever beam structure as serial links; therefore, if the DOF of the master arm increases, the length of the cantilever beam and the total weight of the actuators also increase, thereby decreasing the rigidity and stability of the master arm. Since the measurable movement of the master arm that follows the operator's hand has six DOFs, a new lightweight posture sensor composed of an acceleration sensor is used (Nakagawara *et al.*, 2004) for measuring the final DOF, which is critical to identifying the posture of the operator's entire arm. Altogether, the master arm serves as a master system with seven DOFs for the measurement of the arm's posture and six DOFs for force presentation. Since the posture sensor is very lightweight compared with the mechanical restraints on the operator's elbow, the sensor enables relatively unhindered movement of the operator's arm without any undesirable load on it.

An exoskeleton structure is widely used for master arms since it can be adopted for movement by the operator with minimal size requirements, which is an essential requirement for correspondence to the various everyday actions of humans. The structure and the general appearance of the master arm are shown in Fig. 6.19.

A potentiometer and an encoder are installed in each joint, and the operator's initial posture is computed from the output signals of the potentiometers. During movement, the joint angles and the angular velocities of the operator's arms are computed from the output signals of the encoders. The three axes of the joints in the master arm's wrist cross at

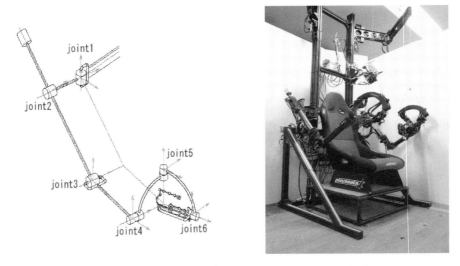

Fig. 6.19. Master arm (left: structure; right: overview).

one point and a 6-axis force sensor (MINI 4/20, BL AUTOTEC) is attached to that point. The output signals of this sensor are used for measuring the force acting between the wrists of the master arm and the operator in the direction of the rotating axis, and the torque that acts around the axis. A bilateral system is realized by attaching the exoskeleton-type multi-fingered master hand to the tip of the master arm.

A gravity compensation system is realized by suspending a wire at the tip of the master arm, which ensures maximum performance of the manipulator. The tension of the wire is 20.6 N. The use of this gravity compensation system means that the actuators of the joints of the master arm do not have to compensate for the gravitational torque applied to the master arm. This makes it possible for the master arm to present forces to the operator's hand with a small output torque and high accuracy. Attached above the master arm is a passive link, which has two joints, and a constant force spring runs through the link. Constant tension acts on the wire by passing it through a pulley at the tip of the spring. The wire is attached to the tip of the master arm. Since the joints of this link are parallel to the direction of gravity, the link can follow the master arm smoothly by means of the wire, which runs through the passive link while maintaining a horizontal posture.

The maximum force at the end of the master arm of 239 N is thus attained.

The master arm has six DOFs and the slave arm has seven DOFs; hence, a simple symmetric servo cannot be employed between corresponding joints. Two requirements are precise hand movements and communication by gestures. In order to satisfy the first requirement, the position and the orientation of the slave arm's wrist must coincide with those of the master arm. Six DOFs are used for this purpose. In order to satisfy the second requirement, the posture of the slave arm must be as close as possible to that of the operator. The remaining one DOF is used for this purpose. It should be noted that it is not appropriate to employ the conventional method based on the pseudoinverse of the Jacobian matrix since it does not satisfy the second requirement of communication by gestures.

A few techniques are available for measuring the operator's posture and include the use of markers and cameras, which are adopted by general motion capture systems. However, the optical method suffers from problems of time delay and occlusion. Therefore, a simpler method outlined below which avoids these problems is considered. Such a method measures the operator's posture using seven DOFs for position and orientation and six DOFs for force feedback on the master side.

As the operator's arm can be regarded as a redundant manipulator, conventional methods to solve the inverse kinematics of a redundant arm can be applied to measure the operator's posture. One popular method is to define the swivel angle of an arc of a circle, which the elbow traces; this angle lies on a plane whose normal is parallel to the wrist-to-shoulder axis. Given the wrist position, orientation, and elbow swivel angle, an algorithm can compute the joint angles analytically. It should be noted that the wrist position and the orientation of the operator are identical to those of the master manipulator.

In order to acquire the remaining information (swivel angle), an acceleration sensor (ADXL202E, Analog Devices, Inc.) is attached to the operator's upper arm. The sensor has a suitable high-frequency response (1 kHz), while its small size and low weight permit the operator to move his or her arm freely. An axis of the sensor corresponding to change in the swivel angle is used for the remaining information, as shown in Fig. 6.20.

6.3.2.2. *Master Hand*

A new type of master hand has been developed, as shown in Fig. 6.21. It has two distinct features. The first is the compact exoskeleton mechanism of the master hand's finger contrived to cover the wide workspace of an operator.

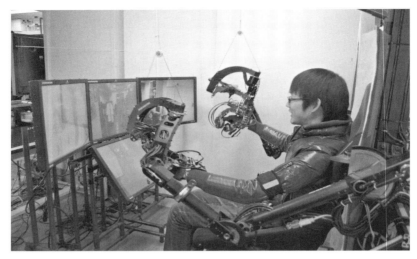

Fig. 6.20.   Acceleration sensor attached to the operator's upper arm along a swivel axis.

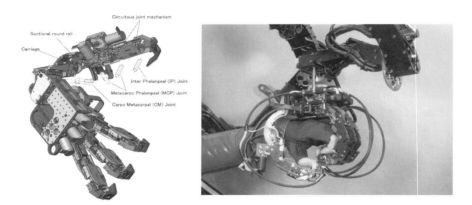

Fig. 6.21.   Master hand (right: structure; left: overview).

The exoskeleton mechanism can be placed either over (parallel joint) or beside (coaxial joint) the operator's finger. The former placement has a disadvantage in that the master arm's finger obstructs the motion of the operator's finger when the operator's finger is bent. However, the latter placement is difficult because there is little space to place the exoskeleton mechanism. To solve this problem, a "circuitous joint" method is proposed (Kawabuchi *et al.*, 2003; Nakagawara *et al.*, 2005). In the "circuitous joint," the joint axis of the master hand coincides with that of the operator by extending the link length in proportion to the joint angular displacement, as shown in Fig. 6.22.

The second feature of our master hand is the encounter-type force feedback (Tachi *et al.*, 1994; Nakagawara *et al.*, 2005). An encounter-type device remains at the location of the object in the remote environment and waits for an operator to encounter it. As shown in Fig. 6.23, our

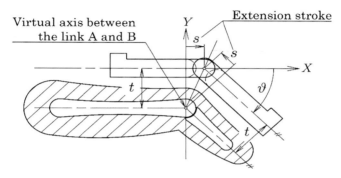

Fig. 6.22.   Basic schematic of the circuitous joint.

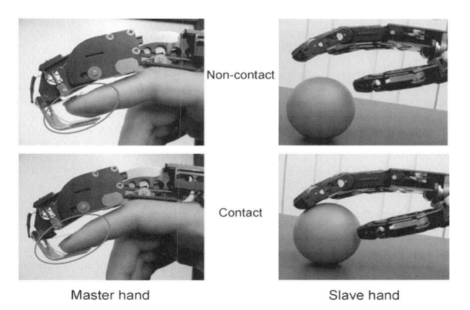

Fig. 6.23.   Encounter-type master–slave hand system.

encounter-type master hand's finger usually follows the operator's finger without physical contact. It enables the operator to touch nothing when the slave hand does not touch anything. When the slave hand touches an object, the master finger stops its movement so that the operator's finger touches a plate of the master hand. The plate provides both the feeling of contact and an appropriate resistive force, thus providing both unconstrained motion and a natural sensation of touch.

Haptic sensation must be presented on the basis of the characteristics of human perception as kinesthetic (force) sensation and tactile sensation. Kinesthetic sensation has essentially the same meaning as proprioception, whereas tactile sensation means cutaneous sensation. Our encounter-type force feedback mentioned above was accomplished using kinesthetic sensation.

In order to present tactile sensation, we mounted an electrotactile display (Kajimoto *et al.*, 2004) on the master hand (Fig. 6.24). This display is a tactile device, which directly activates nerve fibers within the skin surface by an electrical current produced by surface electrodes. It can selectively stimulate each type of receptor and produce vibratory and pressure sensations with an arbitrary frequency.

By using the above-mentioned devices, a master–slave system for haptic telexistence has been constructed. The master–slave manipulation

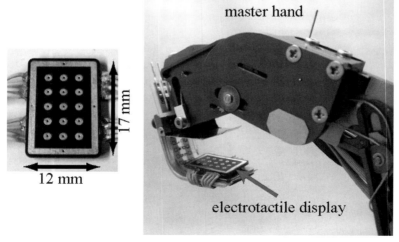

Fig. 6.24. Electrotactile display (left: electrotactile display with 15 electrodes; right: electrotactile display mounted on the master hand).

is realized by bilateral impedance control of the dexterous slave hand and the encounter-type master hand. This control is done from the position of the master and slave fingers and the force applied to them. The position is calculated from the angle of each finger joint, and the force is measured by the finger-shaped GelForce on the slave hand and the force sensor on the master hand. The refresh rate of the control is 1 kHz. Thus, operation is performed smoothly and sufficient kinesthetic sensation is perceived.

When the slave hand touches an object, the finger-shaped GelForce mounted on the slave hand acquires haptic information such as the distribution of the magnitude and the direction of the force. This information is transmitted to the master system in turn. The electrotactile display presents us with tactile sensation based on this information. Using the force distribution, the electrostimuli present information regarding location, and subsequently, the force magnitude information at each position is presented by the strength of the electrostimuli. As a result, we can detect the location, edges, contours, and movement of an object.

By integrating these kinesthetic and tactile sensations as haptic sensation, we can perceive the exact shape and stiffness of an object (Sato *et al.*, 2007). This enables highly realistic interactions with objects in remote places (Fig. 6.25).

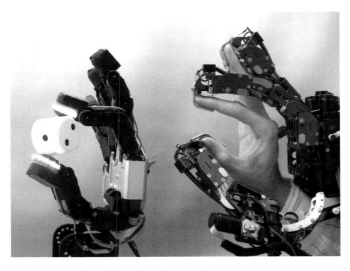

Fig. 6.25. Haptic telexistence system.

6.3.2.3. *3D Display System*

If a head-mounted display (HMD) is used for displaying the 3D remote environment where a surrogate robot is working, people inevitably see an operator with an HMD, which is not preferable from the viewpoint of the desired face-to-face communication.

In light of this, an autostereoscopic 3D display system consisting of four 3D displays (SynthaGram 204: 20 inch lenticular-type LCD display) has been constructed and used. Four displays are arranged underneath and in front and to the left and right of the operator, thereby forming a T-shape. Owing to the placement of the lenticular lenses on the display surface of the LCD display, the operator can view a stereoscopic image without wearing any special glasses, such as shutter or polarized glasses.

The 3D camera system is located on top of the robot. Since the display system is fixed, the camera system should also be fixed. The 3D camera system consists of four pairs of CCDs (eight CCDs) (Tachi *et al.*, 2001, 2003). The pairs of cameras are for the front, right, left, and bottom views. Each image captured by these cameras is transferred to its respective display. The system provides an approximated egocentric view of the operator.

6.3.3. *RPT Viewer System*

As described in Chap. 5, a new type of visual display termed an RPT display is being developed in the author's laboratory at the University of Tokyo and Keio University. The display uses a retroreflective material as its projection surface, such that it functions as a special screen. In the RPT configuration, a projector is arranged at the axial symmetric position of the operator's eyes with reference to a half mirror, with a pinhole placed in front of the projector to ensure an adequate depth of focus. Figure 6.26 illustrates the general appearance of the RPT system used in this study.

The face and the chest of Telesar II are covered by a retroreflective material. A ray incident from a particular direction is reflected in the same direction from the surface of the retroreflective material. This characteristic of the retroreflective material means that an image is projected without distortion onto the surface of Telesar II. Since many RPT projectors are used in different directions and different images are projected corresponding to the cameras placed around the operator, the corresponding images of the operator can be viewed.

Figure 6.27 presents an example of the projected images of an operator onto a surrogate robot.

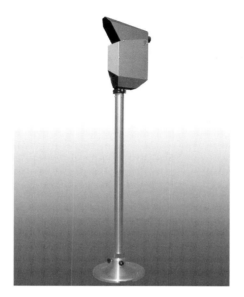

Fig. 6.26.   General appearance of RPT system.

Fig. 6.27.   Examples of RPT appearance (left: TELESAR II without projection; center: with projection of an operator; right: operator at the controls).

6.3.4. *Feasibility Experiments*

In order to demonstrate the feasibility of the concept of mutual telexistence using RPT, the author and his team constructed a hardware system and exhibited it at Expo 2005 Aichi Japan as one of the prototype robots at the Morizo and Kiccoro Exhibition Center from 9 June through 16 June (Kawakami *et al.*, 2005). We constructed two booths: a cockpit booth and a robot booth (Fig. 6.28). We assumed that the cockpit booth was located in

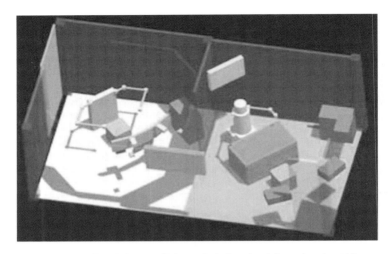

Fig. 6.28.   Booth layout (left: cockpit booth; right: robot booth).

Tokyo and the robot booth, which was a store specializing in selling stuffed toys, was located in Paris. The store had a telexistence communication robot in order to greet foreign customers without them needing to physically travel to the store. A person in the robot booth, which was supposed to be located on a street in Tokyo as a future extension of a telephone booth, could visit the store in Paris by using the telexistence communication system.

Figure 6.29 shows the general appearance of the cockpit booth (left), the robot booth (center), and a customer telexisting using the robot and a clerk (right).

The person (operator) would sit down and wear the master arms and hands to log into the surrogate robot in Paris through a dedicated network. The robot would move its arms and hands corresponding to the operator's motion, and the autostereoscopic display allowed the operator a 3D view of the shop. The operator could communicate with the clerk in the shop using a headset comprising a microphone and a speaker and by using the master–slave communication system to shake hands or make gestures. The operator could not only select a stuffed toy of his or her choice but also handle the merchandise using the master–slave manipulation system. In this manner, a person in Tokyo could feel that he or she was visiting a store in Paris and select merchandise to buy. The clerk in Paris could also see the visitor in the shop by using the RPT viewer system. Figure 6.30 shows an example of such views.

Fig. 6.29. General appearance of the booths (left: cockpit booth; center: robot booth; right: human–robot communication).

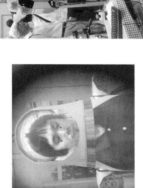

Fig. 6.30. Mutual telexistence using RPT (left: without RPT viewer; center: through RPT viewer; right: observer viewing through RPT).

At the Aichi Expo, we set up three RPT viewers in front, to the left and to the right of the booth. Observers could simultaneously see the front, left side, and right side view of the operator's face. Three cameras were also arranged corresponding to the RPT viewers, allowing three people to see the operator's images simultaneously as projected onto the surrogate robot. One person faced the front view of the operator's face, whereas the remaining two people diagonally faced either the left or right side of the operator's face enabling them to see the corresponding image of the operator. As the operator turned his or her face, he or she could communicate face to face with one of the three people located remotely.

The main features of the proposed system are as follows: (1) conventional verbal communication as well as non-verbal communication is made possible and (2) face-to-face communication is possible under the condition that several people are present.

In order to evaluate the above-mentioned features of the proposed system, the following additional experiments were conducted.

(1) *Non-verbal communication using hand gestures and handshakes*: Several hand gestures such as pointing gestures, iconic gestures, and emblematic gestures including thumbs up and OK were made easily using

Fig. 6.31.   Hand gestures.

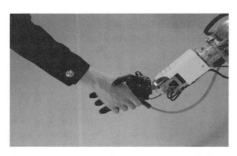

Fig. 6.32.   Handshaking experiment (left: shaking the slave hand; right: operator with the master hand).

the developed master–slave system. Figure 6.31 shows some examples of the realization of such gestures.

A handshake action was performed successfully, with the positions and forces of the slave and master arms measured. Figure 6.32 shows an example of a handshake with a surrogate robot.

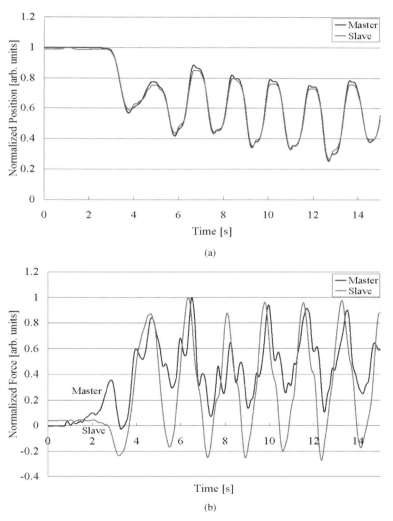

Fig. 6.33. (a) Normalized positions and (b) normalized forces of slave and master during a handshake.

Figures 6.33(a) and 6.33(b) show examples of the measured position and force in the respective upward and downward directions of the handshake. As shown in this figure, motion is initiated by the slave since the action of a handshake with a surrogate robot requires initiative on the part of the person shaking hands with the slave. The graph shows that the position of the master follows the position of the slave without any delay and the force of the master is delayed compared with the force of the slave by around 0.2 s. At the beginning of the handshake action, antagonistic force is produced on the master side and no movement occurs for approximately 3 s before the person on the master side follows the motion of the person on the slave side. These movements were clearly recorded and are shown in Fig. 6.33.

(2) *Estimation of the maximum number of people who can view the operator's image simultaneously:* Figure 6.34 shows the characteristics of the retroreflective material used to cover the surface of the robot (Ref-lite). A reflectance of 100% with reference to normal incidence is obtained for an incident angle of more than 60°, which is sufficient for a human-sized robot.

Figure 6.35 shows an example of a projected image captured from several positions around the robot. Clear images of the operator are obtained from any angle because of the reflectance characteristics shown in Fig. 6.34.

In order to estimate the maximum number of people able to view the operator's image simultaneously, the following measurement was performed. A white uniform image was projected onto the retroreflective surface of the robot, and reflectance was measured as a function of the angle between the

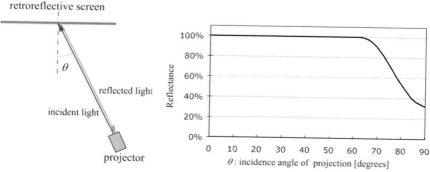

Fig. 6.34. Reflectance of retroreflective material as a function of the incidence angle of projection.

Fig. 6.35. Projected images of the operator captured from various angles.

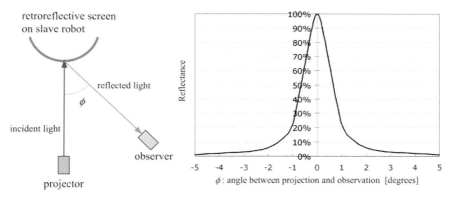

Fig. 6.36. Reflectance of retroreflective material as a function of the angle between projection and observation.

projector and an observer, as shown in the left panel of Fig. 6.36. The result is shown in the right panel and indicates that virtually all light is reflected within an angle of 3°, implying that when we separate two observation points at an angle of 3°, no interference occurs. If we simply divide the view of 360° by 3°, 120 simultaneous viewing points are obtained.

6.3.5. *Summary*

The concept of using a robotic mutual telexistence system for natural face-to-face communication was proposed, and its feasibility was demonstrated by constructing a mutual telexistence master–slave system using RPT.

For face-to-face communication to occur between two people in different locations, person "A" must be able to see another person "B" face to face, and *vice versa*. The author proposed a system in which person "A" uses a surrogate robot located in the place where person "B" is present and the surrogate robot is covered with retroreflective material so that a real-time image of "A" could be projected onto it. It was demonstrated experimentally by using the proposed RPT-based mutual telexistence method that not only was "A" able to see "B" but also that "B" was able to see "A" face to face as if they were actually face to face.

It was also proven that non-verbal communication such as gestures and handshakes could be performed in addition to conventional verbal communication with the use of a master–slave manipulation robot as the human surrogate.

Finally, it was shown that person "A," who "visited" a remote location where a surrogate robot was located, could be seen naturally and simultaneously by several people standing around the surrogate robot.

Chapter 7

Telexistence Communication Using TWISTER: Telexistence Wide-angle Immersive STEReoscope

TWISTER (Telexistence Wide-angle Immersive STEReoscope) is an immersive omnidirectional full-color autostereoscopic 3D booth, designed for a concept of face-to-face telecommunication referred to as "mutual telexistence," where people at distant locations can communicate as if they were present in the same 3D space. Each user is situated within a cylindrical booth, which displays full-color 360° panoramic and autostereoscopic images in real time to the user without the need for any eyewear. At the same time, the booth captures images of the user from every angle. By using multiple booths, a number of people at different remote locations can interact with each other as if they were interacting face to face.

This concept was first proposed in 1996, and prototype models TWISTER I, II, III, IV, and V were constructed based on this concept. TWISTER IV displays binocular stereo images using 36 display units, each consisting of two LED arrays for the left and right eyes and a parallax barrier that rotates at high speed (1.7 rps). This system simultaneously captures 3D images of the observer by employing cameras installed on the rotating cylinder. Thus, observers inside two separate TWISTERs can view 3D images of each other in real time. This chapter explains the concept and principles of TWISTER in relation to mutual telexistence, and describes the prototype systems developed and possible applications of TWISTER.

7.1. Face-to-Face Communication

Face-to-face communication is an essential and inevitable component of everyday life. However, in situations where people are far apart, a face-to-face telecommunication system proves useful. One of the most essential elements of such a system is providing a sensation of presence to the users. Immersion, as one of the essential factors for creating a sensation of presence, is desirable in that the visual display of such a system provides an immersive stereoscopic view. Two typical conventional immersive systems include a head-mounted display (HMD) and CAVE (CAVE automatic virtual environment). In such conventional systems, however, the mechanisms for binocular stereopsis, such as the headgear required for an HMD and shutter-glasses for CAVE, not only leave the observer with a sense of discomfort but also hinder face-to-face communication since they obscure the observer's face, particularly around the eyes.

Autostereoscopic 3D displays are effective for face-to-face communication because they enable the user to view 3D images without the use of any special headgear or eyeglasses. The most common autostereoscopic 3D displays employ lenticular lenses or parallax barriers. Using such displays, if a viewer positions his or her head in a certain manner, he or she will perceive different images with each eye, which gives rise to a stereo image, but a 360° panoramic field of view cannot be realized. While integral imaging using fly's eye lenses is a promising technology, it has not as yet been employed in any 360° integral imaging autostereoscopic 3D display.

The conventional parallax barriers, lenticular lenses, or fly's eye lenses have another limitation: it is considerably difficult to take a picture of a user's face while he or she is observing 3D images using these systems. This limitation stems from the fact that the screen is too densely packed with microlenses and LCDs to locate cameras in between. Since recording a user's face in real time at eye level is essential for face-to-face telecommunication, the cameras should be placed at the level of the user's eyes. However, such a setup leads to sparse arrangement of microlenses, which adversely affects the performance of the display.

In order to resolve such problems, the TWISTER is proposed as a booth for face-to-face telecommunication. By adopting the moving parallax barrier method, TWISTER can display panoramic stereoscopic images that can be observed without the use of any special eyewear. Moreover, we can place cameras between the display units. By rotating the cameras simultaneously, the booth can display stereoscopic images and capture the figure of the user simultaneously. It is also possible to place cameras outside the booth and view the user from outside. Once these images are obtained from various viewpoints around the user, a new view from an arbitrary viewpoint can be constructed using an image-based rendering technique.

With the use of several booths, each remote user can view the 3D figures of other users working in real time in the mutual virtual environment. In such a situation, the booth works as the medium between real space and virtual communication space.

In this chapter, the concept, principles, developed prototypes, and possible applications of TWISTER are described.

7.2. Concept of Mutual Telexistence Using TWISTER

A different approach to mutual telexistence without the use of robotics can be taken if we restrict our purposes just to communication. The basic concept is shown in Fig. 7.1. Each user stands inside a booth with a rotating cylindrical mechanism and the booth plays the role of both a display device and an input camera device, which captures moving pictures of the user inside the booth. Each user can see the 3D figures of other users working in real time in the mutual virtual environment. This concept was first proposed in 1996, and a preliminary experiment successfully demonstrated its feasibility (Tachi *et al.*, 1996).

Figure 7.2 illustrates the concept of telexistence in a mutual virtual environment. Each user is situated inside a booth that has a virtual cylindrical wall, which can be used as a display of the virtual environment onto which all users are projected as well as an input camera device to obtain the 3D moving pictures of the user. The booth is surrounded by (1) a circular mechanism that turns continuously at a relatively high-speed

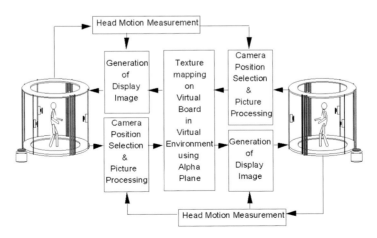

Fig. 7.1.   Basic concept of the mutual telexistence booth.

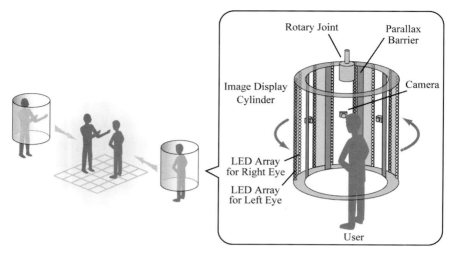

Fig. 7.2.   Concept of mutual telexistence.

carrying display units consisting of a pair of LED arrays and a parallax
barrier and (2) video cameras.

The video cameras, which are arranged such that they cover the user completely, take pictures of the user from virtually every angle as they turn around
the user inside the booth. Using this method, stereo information pertaining to
the user can be obtained from virtually any angle. When the system is used
by several users, each user's line of sight and relative position are measured
and are used to select the appropriate picture of other users. The images of
other persons thus selected are appropriately arranged three-dimensionally in
a mutual environment and displayed to all the users.

Virtual transparent boards (alpha plane) are placed in a virtual
environment, and the pictures taken from two positions (left and right)
are texture mapped on the boards. The other parts of the scenery are not
texture mapped and thus remain transparent. The corresponding points of
the texture-mapped human figures on the virtual boards produce the 3D
figure of the user.

7.3. Development of TWISTER

As mentioned in the previous section, the idea of mutual telexistence using
cylindrical booths that can be used for both a panoramic display and a real-
time camera system, which takes pictures of a user from any direction, was
initially proposed in 1996 (Tachi *et al.*, 1996). The handmade experimental

system constructed, in which the functions of the input device and display device were separated for the sake of simplicity, demonstrated the feasibility of the idea. Fundamental studies aimed at realizing the two functions of input and display were subsequently conducted, and thereafter the first handmade prototype system with display function, dubbed TWISTER I, was constructed in 2000, and the results were reported in 2001 (Kunita *et al.*, 2001).

The CREST project on Telexistence Communication Systems led by the author and funded by the Japan Science and Technology Agency (JST) commenced in November 2000 and ended in March 2006 (Tachi, 2003a, 2005). Under this project, the concept of TWISTER was applied to a tangible real system that can be experienced.

TWISTER II was constructed on the basis of TWISTER I and reported in 2001 (Tanaka *et al.*, 2001). TWISTER II successfully realized the world's first omnidirectional autostereoscopic full-color display. It was demonstrated to the world at SIGGRAPH in 2002 using TWISTER III (Tanaka *et al.*, 2002, 2004). TWISTER IV is a system with functions of both display and video input (Hayashi *et al.*, 2002) and was constructed in 2004. A very similar system called TWISTER V was constructed in 2005 and had a place in the National Museum of Emerging Science and Innovation (Miraikan). Communication experiments are currently being conducted using TWISTER IV (at the University of Tokyo) and TWISTER V (at Keio University).

Figures 7.3 and 7.4 show the typical panoramic images displayed in TWISTER.

Table 7.1 shows the specifications of TWISTER III and TWISTER IV. Incidentally, the specification for TWISTER V is essentially the same as that of TWISTER IV; the difference between them is that minor modification of the LED arrangements on a display unit was made for TWISTER V so that the LED display can be used without a diffuser.

7.4. Principles of TWISTER

7.4.1. *Movable Parallax Barrier*

Movable parallax barriers constitute the key device for autostereopsis. Figure 7.5 shows the principle of the movable (rotating) parallax barriers developed. Each display unit contains two LED arrays — one for the left eye, and the other for the right eye. The units rotate at a speed of approximately 1.7 revolutions per second (rps), and the controller synchronizes the display update to create an effective frame rate of 60 frames per second (fps).

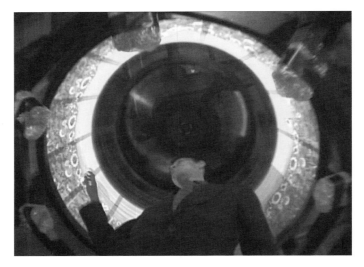

Fig. 7.3.   360° panoramic image displayed in TWISTER II.

Fig. 7.4.   360° panoramic image displayed in TWISTER IV.

Since the parallax barrier obscures LED emission from the opposite side, different images are shown to the left and right eyes. At one moment, only images for the left eye and right eye are observed in areas L and R, respectively. Both images are observed in area B, while neither image is

Table 7.1. Specifications of TWISTER III and TWISTER IV.

Specifications	TWISTER III	TWISTER IV
Horizontal resolution (pixel/360°)	1920	3168
Vertical resolution (pixel/74°)	256	600
Radius of the display (mm)	800	1000
Height of the display (mm)	960	1200
Pixel pitch (mm)	3.75	2
Visual acuity	0.06	0.15
Frame rate (fps)	30	60
Luminance gradation	RGB 8 bit × 3	RGB 10 bit × 3
Input format	NTSC(240 × 256) × 8ch	DVI(1600 × 1200) × 2ch
Rotation speed (rps)	1.2	1.7
Number of display unit (units)	30	36
Camera resolution	—	VGA
Radius of camera rotation circle (mm)	—	950
Camera angle of view (°)	—	45
Camera frame rate (fps)	—	30
Camera output format	—	IEEE1394 × 2ch

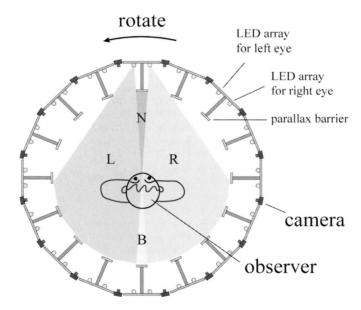

Fig. 7.5. Principle of the movable (rotating) parallax barrier method.

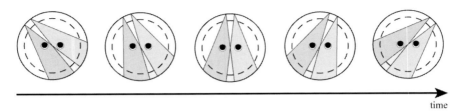

Fig. 7.6.   Parallax barrier rotates maintaining the stereoscopic condition.

observed in area N. When the display units with barriers rotate around the user, these areas also rotate, as shown in Fig. 7.6.

The condition of stereopsis is maintained as the display units rotate around the user. The angle of view of a stereoscopic display depends on the direction and the position of the observer. If the head of the observer is fixed in one position, it exceeds 140° in an ideal condition. On the other hand, if the observer always faces the center of the region of interest of the image, the angle of view can be 360°.

The use of rotating parallax barriers reduces the crosstalk between the left and right eye images to almost zero, which gives this system a powerful advantage over other autostereoscopic vision systems. Moreover, the barrier *per se* cannot be observed by the observer because it rotates at a speed of 600° per second — the human eye can track a revolution of only approximately 500° per second. This assures a continuous 360° panoramic stereo view.

7.4.2. *Rendering*

TWISTER employs two modes in rendering. In Mode I as shown in Fig. 7.7, we assume we can track the direction in which the head of the user is facing and render the left and right images on a display cylinder perspective transformed from the left eye and right eye, respectively. In this mode, a precise image is obtained when the head is fixed in one position. The condition is kept if we render in real time according to the measured head direction.

In Mode II as shown in Fig. 7.8, rendering with concentric mosaics is used. Here, irrespective of the direction the user turns in, correct light information will be obtained at least from the front. The light information from other directions is an approximation. However, it is not necessary to track the direction in which the user's head faces.

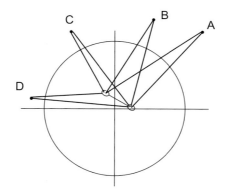

Fig. 7.7.   Rendering Mode I.

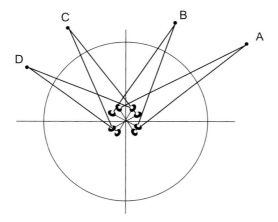

Fig. 7.8.   Rendering Mode II.

Mode II is typically used since it does not require the use of a tracking system, making it a more convenient option.

7.4.3. *Image Capture*

Figure 7.9 shows the principle of the image capture method (Tanaka *et al.*, 2001). To synthesize the situation in Fig. 7.9(a), cameras in booth A capture the light rays as seen from the supposed direction of the observer. These rays are transmitted and displayed on the cylindrical display of booth B (Fig. 7.9(b)).

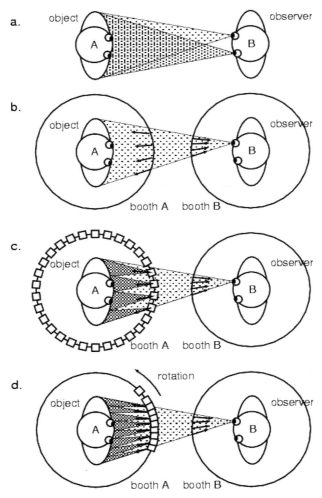

Fig. 7.9.   Principle of capture and display: (a) supposed situation, (b) ray group seen by the observer, (c) ray group captured by cameras, and (d) selection of appropriate rays from those captured by the rotating cameras.

    In general, in order to display 3D objects for an arbitrary viewpoint outside a closed surface, ray information is required from a sufficient variation of directions inward, and at points of sufficient density over a closed surface separating the observer and the object. However, it is difficult to capture images from continuous viewpoints simultaneously since each camera occupies physical space. In order to increase the number of

Fig. 7.10. Constructed human figure. Images from an arbitrary viewpoint are constructed using rays captured by a camera array and placed in a virtual environment.

viewpoints, we considered two approaches: to reduce the camera size (c), and to exploit the camera's motion (d).

Figure 7.10 shows an example of capturing a human figure in real time and displaying it in a virtual environment (Kunita *et al.*, 2001). First, with a virtual viewpoint given by a user, the rendering PC calculates the regions of each camera image to be used for multi-texture mapping. Next, the information of the regions is sent to the control PC and used as an indication of the video switch. The switch timing is synchronized with the camera scanning, and the camera scanning direction is vertical. In this manner, the rendering PC can selectively capture the necessary column image from the cameras.

The captured images are then texture mapped onto a plane at a corresponding focal distance in the virtual 3D space. This process is equivalent to the simple memory copy. Finally, the rendering PC renders the scene with 3D graphics. One cycle of all these processes completes within video rate (1/30 s), and thus the system realizes real-time rendering.

7.4.4. *Capture and Display*

As mentioned above, the use of moving parallax barriers reduces the crosstalk between the left and right eye images to almost zero, providing a powerful advantage over other stereoscopic vision systems. Since the rotating unit has cameras mounted on it, the image of the observer can be captured simultaneously. With this system, the observer's face is clearly captured, and natural non-verbal communication between users in multiple booths is achieved when the image data are transferred in real time.

The rotator rotates counterclockwise (viewed from the top) at a constant speed of about 1.7 rps. The sync signal is generated when a photo detector attached to the rotator senses the light from a photo diode attached to the framework.

Display units are attached at a distance of 1,000 millimeters from the center. Under such a condition, with vergence, the observer is able to fuse an image of an object at a distance ranging from 0.2 m to infinity, which meets the purpose of face-to-face telecommunication.

The stereo video picture of a user inside the booth is captured by 36 rotating cameras, and an image from an assigned angle is reproduced from the information captured by appropriate cameras. Figure 7.11 shows a schematic diagram of the system, and Fig. 7.12 shows a preliminary telexistence communication experiment using TWISTER II.

Although it was proven that the proposed method can capture a user's stereo video picture from an arbitrary direction by employing rotating cameras inside the TWISTER IV and V booths (Hayashi *et al.*, 2002), it became apparent that the use of this method is impractical at present

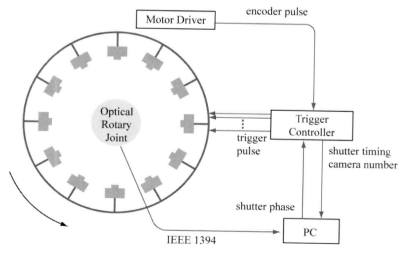

Fig. 7.11. Schematic diagram of the human stereo video picture capturing system.

Fig. 7.12. Preliminary experimental telexistence communication using TWISTER II.

because it requires

(1) an expensive high-speed large-capacity rotary joint, and
(2) precise alignment and calibration of video cameras.

Thus, an alternative method of image capturing from an arbitrary direction is proposed. As clearly seen from Figs. 7.13 and 7.14, one of the salient characteristics of TWISTER is that the user inside the booth can

Fig. 7.13. User observed from outside the TWISTER II booth.

Fig. 7.14.   User observed from outside the TWISTER IV booth.

be seen from outside the booth. This implies that the human user's picture can be taken from any direction from outside of the booth.

Figure 7.15 shows a schematic diagram of the outside-camera method. A movable stereo camera is set on a concentric rail arranged outside the booth and the stereo camera is controlled in accordance with the angle from which the human picture should be taken. Stereo cameras can be used in the case that several people need to observe the user inside the booth. When it is necessary to cross two cameras, they can virtually be crossed by changing the roll of both cameras.

Figure 7.16 shows an implementation of TWISTER for face-to-face communication in virtual space. Consider the situation shown in Fig. 7.16. Three users are situated inside different TWISTER booths. Each booth has two stereo cameras. Although the human users are in remote locations, they can meet in a virtual environment using the TWISTER system.

First, a 360° 3D virtual environment is displayed on each booth. Next, the two stereo cameras in each booth capture pictures of the user inside the booth from the angles calculated on the basis of the relative position and orientation of the three users in their virtual environments. The captured images are then transmitted using communication links to the

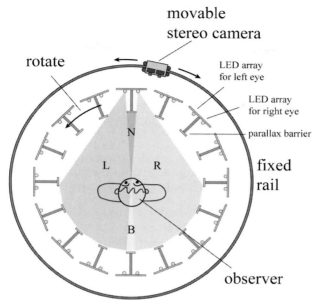

Fig. 7.15.   Method of capturing human stereo video picture from outside the TWISTER booth.

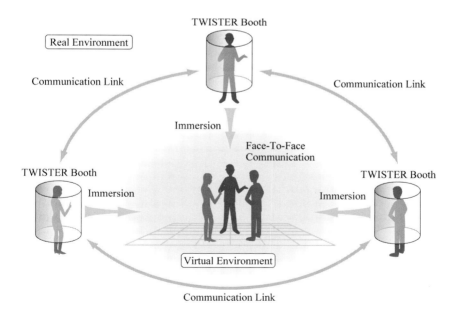

Fig. 7.16.   Face-to-face communication using TWISTER booths.

other TWISTERs. Inside each booth, the figures of the other two users are placed in the virtual environment according to their relative position and orientation. In this manner, the users sense that they are in close proximity in the virtual environment.

Technology based on the concept of networked telexistence enables users to meet and talk as if they share the same space and time, even if they are located remotely. This is the goal of developing mutual telexistence communication systems and is a natural progression from telephone to telexistence-videophone.

7.4.5. *Robotic Mutual Telexistence Using TWISTER*

Mutual telexistence is one of the most important technologies for the realization of networked telexistence since users "telexisting" in a robot must know with whom they are working over the network, as explained in Chap. 6. Figure 7.17 shows how TWISTER is used as a cockpit for mutual telexistence.

User A can observe remote environment [B] using an omnistereo camera on a surrogate robot A'. This provides user A with a panoramic stereo view of the remote environment displayed inside the TWISTER booth. User

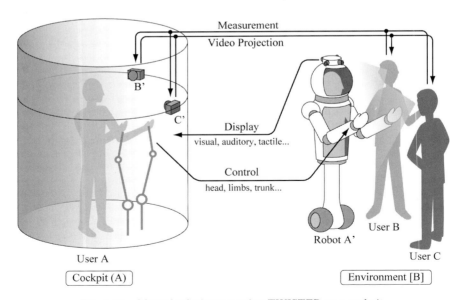

Fig. 7.17.   Mutual telexistence using TWISTER as a cockpit.

A controls robot A′ by using the telexistence master–slave control method described in Chap. 6.

Cameras B′ and C′ mounted on the booth are controlled by the position and orientation of users B and C relative to robot A′, respectively. Users B and C can observe the image of user A projected onto robot A′ by using a head-mounted projector, as described in Chap. 5. Since robot A′ is covered with retroreflective material, it is possible to project images from both cameras B′ and C′ onto the same robot while still having both images viewed separately by users B and C.

This method of mutual telexistence using retroreflective projection technology (RPT) was proven to be effective, as discussed in detail in Chap. 6.

Figure 7.18 shows an example of the projection of a user's figure onto a robot.

7.4.6. *Omnistereo Camera System for TWISTER*

In order to complete the mutual telexistence system described in Sec. 7.4.5, it is necessary to develop an omnistereo camera system for TWISTER.

Figure 7.19 (left) shows the camera configuration corresponding to the left eye of an observer during omnistereo imaging. When the observer turns his or her head by $\theta°$, the left eye (or the camera) moves and the light axis is rotated. In other words, the position of each camera varies according to the viewing direction. Adding this one dimension to the other two dimensions of the captured image, total 3D ray information has to be recorded at every instant. Here, we can reduce one dimension by replacing the normal

Fig. 7.18. Example of the projection of a human figure on the Telesar II robot.

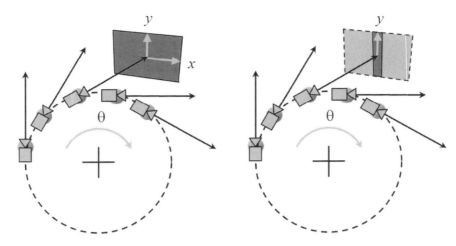

Fig. 7.19. Formulation of omnistereo imaging.

camera with a line scan camera, as shown in Fig. 7.19 (right). With this approximation, the generated image is always accurate with regard to the observer's position facing frontward, even if his or her head rotates, and the incoming light rays from other directions are approximated by the perpendicular incident light rays to his or her left (or right) eye when his or her head turns in another direction. We refer to this 2D light ray expression (θ and y) as omnistereo imaging.

Omnistereo imaging is a subset of concentric mosaics where the radius parameter from three parameters of the concentric mosaics is fixed at half the interpupillary distance (IPD). Figure 7.20 is an example of an omnistereo image constructed by panning two relatively fixed line scan

Fig. 7.20. An example of an omnistereo image.

cameras. Points at infinity are plotted in the same horizontal positions in the left and right images.

A practical method for capturing omnistereo video sequences using rotating optics has been proposed and evaluated (Tanaka and Tachi, 2005). The rotating optics system consists of prism sheets, circular or linear polarizing films, and a hyperboloidal mirror. This system has two different modes of operation with regard to the separation of images for the left and right eyes. In the high-speed shutter mode, images are separated using post-image processing, while in the low-speed shutter mode, the image separation is completed by optics. Figure 7.21 shows the whole structure of the system in the low-speed shutter mode. Here, light rays reflected by the hyperboloidal mirror are split by the beam splitter into two groups of

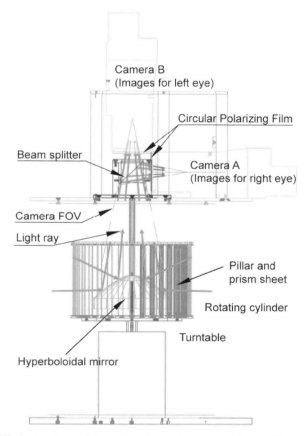

Fig. 7.21. Whole structure of the system for capturing omnistereo video sequences using rotating optics in the low-speed shutter mode.

Fig. 7.22.   Capture of omnistereo video sequences using low-speed shutter mode.

light rays with opposite-hand circular polarization. The left-eye and right-eye light rays are extracted from each group of light rays by right-handed circular polarizing films.

Figure 7.22 shows a general view of an experiment using a constructed omnistereo device referred to as TORNADO, and Fig. 7.23 shows the result.

Fig. 7.23.   Example of omnistereo images.

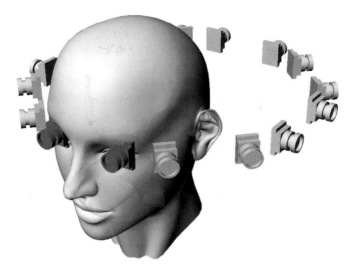

Fig. 7.24. Approximation of omnistereo using fixed cameras.

Fig. 7.25. Constructed omnistereo camera.

Through the capture of actual images, the effectiveness of the methods was confirmed.

Figure 7.24 shows an approximation of omnistereo imaging using a fixed finite number of cameras, and Fig. 7.25 shows such an implementation using 16 cameras. Figure 7.26 presents an example of the omnistereo images

Fig. 7.26. Example of omnistereo images.

obtained. The effectiveness of using approximated omnistereo cameras was clearly demonstrated (Kato *et al.*, 2008).

7.5. Summary

TWISTER was developed for the concept of face-to-face telecommunication referred to as "mutual telexistence," and its efficacy was demonstrated and proven experimentally. It is expected that it will be adopted as a telexistence booth for communication and control that offers users a sensation of presence, and will be widely used with several applications in the 3Cs (communication, control, creation) and the 3Es (experience/education, entertainment, and elucidation). In short, TWISTER is a media booth and it can be used as a future telephone, a future computer display, a future TV, and a future control cockpit for telexistence robots.

Chapter 8

Future Perspective

8.1. Out-of-the-Body Experience

Telexistence is fundamentally a technology that enables a human being to experience a real-time sensation of being in a place that is at some distance from his or her actual location. Furthermore, he or she can interact with a remote environment that may be real, virtual, or a combination of both. The author believes that telexistence has the potential to release human beings from the restrictions of their cognitive limits and physical constraints. It provides a human being with an out-of-the-body experience.

I can never forget the wonderment of undergoing an out-of-the-body experience for the first time when I constructed a telexistence machine and saw myself through it in 1981. I remember that I was looking at my own back, who was observing someone, and I asked myself if I was actually looking at my own self. Then, who was I, who was looking at my own self?

Susumu Tachi,
1981

*It was as if I was standing a few feet away **in another body** looking at myself. I moved my head to look up and down and even to look away. And when I looked away from that person who was me, it was as if that body were just another passerby. But I could not ignore that guy for long, and I turned my head back to look again at this person who was standing across from me with his head strapped into a black-velvet box. The scientists in the laboratory laughed. They knew what was going through my head, for it had also gone through theirs during their own encounters with their out-of-body* *selves. "Are you here?" Tachi laughed. "Or are you there? Where is your body?"*

Grant Fjermedal,
"The Tomorrow Makers," 1986

*The strangest moment was when Dr. Tachi told me to look to my right. There was a guy in a dark blue suit and light blue painted shoes reclining in a dentist's chair. He was looking to his right, so I could see the bald spot on the back of his head. He looked like me, and abstractedly I understood that he **was** me, but I know who me is, and me is **here**. He, on the other hand, **was there**. It doesn't take a high degree of sensory verisimilitude to create a sense of remote presence. The fact that the goniometer and the control computer made for very close coupling between my movements and the robot's movements was more important than high-resolution video or 3D audio. It was an out-of-the-body experience, no doubt about it.*

Howard Rheingold,
"Virtual Reality," 1991

Telexistence research is expected to enhance human life via the extension of human cognition-based behavior and the liberation of human beings from spatial restrictions, and through this, it addresses the concept of human cognition of time and space. With the help of telexistence technology, the author hopes to extend human cognition-based behavior by changing the manner in which humans perceive time and space. Science and technology have always pressed mankind to change the concepts of time and space. Telexistence research can cause another big revolution in this field.

Furthermore, telexistence research is expected to aid the provision of answers to questions such as what a human being is and what self-identity is, by allowing humans to objectively observe themselves from another place through their alter ego. Using this technology, we can observe ourselves in real time, as we would observe our image in a mirror, but from arbitrary viewpoints. Consequently, we can achieve externality with respect to not only visual but also auditory and tactile sensations. This would allow people to recognize themselves objectively and release human cognition from the current cognition frame. Telexistence provides us with a "new mirror";

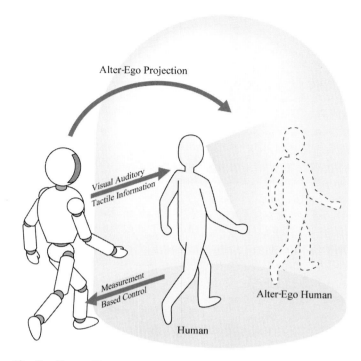

Fig. 8.1.   Alter-ego projection.

further, science would have a powerful tool in the form of telexistence for questioning our identity or the notion of self (Fig. 8.1).

8.2. Impact of Telexistence on Daily Life

Telexistence has the power to dramatically change our concept of space. Through the construction of a room in which a unified transmission of people and objects is enabled using telexistence technology, innovations can be expected in the field of communication technology, in manufacturing enterprises whose production sites are distributed overseas, and even in medical care systems of remote islands. Further, telexistence will have an impact on the development of new mail-order businesses, virtual tourist industries, and urban architecture. All these developments will lead to a major transformation in human life. The features of telexistence technology are different from those of 2D or traditional 3D technologies:

(1) Telexistence technology transmits and reproduces a life-sized 3D space with a realistic feeling of existence.
(2) It enables real-time interaction in the transmitted space.
(3) It enables self-projection of a human in the transmitted space.

Because of these features, the following realizations would be possible in the real world (Fig. 8.2):

- If telexistence booths substitute telephone booths on streets or in offices, many people could meet each other using these booths, as if they were seeing each other face to face. Such technology could support not only conferences but also situations wherein many people walk around freely, such as at cocktail parties.
- For the purpose of easy management, companies whose production sites are distributed overseas could link their branch offices and factories and treat them as though they were virtually located at one place.
- Advanced medical care could be provided worldwide by employing such a transmission system in operating rooms.
- The virtual presence of doctors would facilitate the provision of emergency medical services at early stages of an illness.
- Even at the time of disasters, search and rescue operations would be possible without any concerns for ensuing disasters.
- One could immediately return from a remote place to one's office or home with a real-time sensation of existence.
- One could meet with family, relatives, and friends at remote places in a manner similar to a face-to-face meeting.
- One could have special experiences such as being in the midst of a flock of birds by using a small robot bird as his or her surrogate.
- Life-altering experiences such as looking at the earth from outer space, which is now physically possible only for astronauts, could be offered to everyone.
- One could enjoy real shopping at a remote shop as if he or she were physically present at the shop.
- One could travel to seldom-visited places with the sensation of existence at those places.
- Life-sized space designs and evaluation tools would be available to architects.

Fig. 8.2.   Examples of future applications of telexistence.

Virtual reality is also effective from the viewpoints of a low impact on the global environment and respect for individuals, for the following reasons:

(1) Since products can be manufactured and evaluated in a computer-generated space, manufacture of intermediate waste products and the use of unnecessary energy or materials can be avoided.
(2) Custom-made products designed to suit an individual's needs can be manufactured using a virtually connected space between the producer and the consumer.
(3) Unnecessary energy consumption related to the physical transfer of humans is eliminated.

8.3. Open Problems of Telexistence

(1) Telexistence is a challenging research area that marks the beginning of the true realization of a "holodeck" or a "door to everywhere (dokodemo

door)," as popularly known in science fiction or comics, using available technologies. In other words, telexistence research attempts to utilize somewhat untamed technologies for the purpose of virtual scanning, transmitting and pasting real objects, and making them existent, similar to documents or figures on a computer screen. The virtual reproduction of a physical object (object facsimile) is likely to be one of the open problems of telexistence. This problem includes a virtual cut and copy of a physical object by scanning, presumption of the inner state by scanning of a physical object, addition of attribute information, enhancement of reality, searchable tagging, and virtual pasting of objects by virtual printing (Figs. 8.3–8.5).

(2) The key factor for achieving self-projection is identifying how close the virtual-to-real conversion (VRC) surface should be set to the human body surface at the time of reconstructing an existence field. In order to perceive the alter ego in the existence field as if it were one's own body, it is also important to convey sensory information obtained from the alter ego on the real-to-virtual conversion (RVC) surface to the human body and display it with sufficient accuracy. While some advances have been made in the areas of visual and auditory presentation, no significant advances have been made on the presentation of somatic sensation. The

Fig. 8.3. Virtual holodeck created using telexistence.

Fig. 8.4.   Virtual door to everywhere.

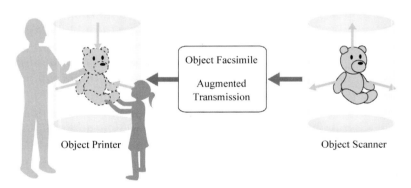

Fig. 8.5.   Object facsimile.

presentation and display of a cutaneous sensation among other somatic senses is extremely challenging and is considered as one of the open problems of telexistence. This problem includes sensing of the existence field on the RVC surface, acquisition of extrasensory information by sensor systems, augmentation using extrasensory information, annotation using knowledge-based information, virtual tagging, reconstruction of existence

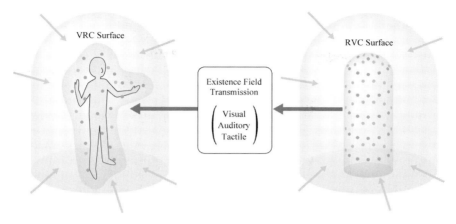

Fig. 8.6.   VRC and RVC surfaces.

fields on the VRC surface, presentation of audiovisual information, and presentation of somatesthesia, tactile or haptic information (Fig. 8.6).

(3) The scientific elucidation of the notion of self is achieved using the existence field technology–human beings recognize themselves through a new mirror. The mirror reflects a person moving freely in a life-sized 3D space not only visually but also auditorily and tactually. It would be a great challenge to establish this approach and use it as a tool for the scientific elucidation of the notion of self. This problem includes human cognition mechanisms for experiencing a realistic sensation of presence, human cognition mechanisms for recognizing the existence of objects, and the human concept of self or identity in the existence field.

8.4. Telexistence in the Future

Telexistence technology has a broad range of applications such as operations in dangerous or poor working conditions within factories, plants, or industrial complexes; maintenance inspections and repairs at atomic power plants; search, repair, and assembly operations in space or the sea; and search and rescue operations for victims of disasters and repair and reconstruction efforts in the aftermath. It also has applications in areas of everyday life, such as garbage collection and scavenging; civil engineering; agriculture, forestry, and fisheries industries; medicine and welfare; policing; exploration; leisure; and substitutes for test pilots and test drivers. These applications of telexistence are inclined toward the concept of conducting operations as a human–machine system, whereby a human compensates for features that are lacking in a robot or machine.

There is also a contrasting perspective on telexistence, according to which it is used to supplement and extend human abilities. This approach involves the personal utilization of telexistence, that is, liberating humans from the bonds of time and space by using telexistence technology.

For example, the president of a company that has branches worldwide cannot make optimal decisions only on the basis of reports presented by branch managers and factory managers on the state of each branch and factory. Rather, it is desirable for the president to personally visit each location and directly observe the conditions in detail. This requires the president to travel extensively, which results in a waste of both time and effort.

Telexistence resolves all issues in such cases. A robot that functions as an alternate identity is placed in each factory or branch office, and whenever required, a communication channel is established with one of these robots so that the robot may be used as an alter ego, and one can see through its eyes, hear by means of its ears, work using its hands, and move using its feet. Videophones provide this functionality in part, but they are passive and their location is restricted.

Using telexistence, one may freely enter and exit any part of a factory, obtain exclusive and accurate sensory information through the sensation of presence, and feel and manipulate objects. In short, through telexistence, it is possible to acquire the same experience as that acquired by actually visiting the location.

Additionally, since the robot acting as one's alternate identity may enter places that are dangerous for humans, in a sense, the function of robots exceeds that of humans. Over and above this, since the robots used in telexistence are intelligent robots, when they are not being used in the telexistence mode, they can act as exceptional assistants and move according to their own judgment, collect information, compile it, and report it to the president. In particular, when the robot encounters something that would be of interest to its master, it communicates with its master. In this case, the master responds to the appeal and utilizes the robot in the telexistence mode, thus solving problems faced during a given situation and gaining experience.

There are many examples of such utilizations of robots. At times, it can be difficult to receive diagnosis and treatment from a specialist doctor, and such attention is highly valuable and hard to relinquish. If a person is granted permission to telexist, he or she can easily make an appointment with a specialist doctor regardless of the location of the hospital where the specialist is present.

Telexistence robots can be set up at hospitals; then, a communication channel can be established with the hospital where a specialist is located, and the robot can be operated in the telexistence mode by the specialist. If the specialist is taken around the hospital and conducts the rounds in full using the telexistence robot, the doctors accompanying the specialist may improve their skills in addition to helping many patients.

Such actions are most likely to cause a dramatic increase in the capability of doctors on a global level. If telexistence is used with robots assigned to ambulances, it is possible for doctors to administer first aid to patients while giving appropriate instructions to ambulance staff on the scene.

Further, telexistence has applications that exploit opportunities provided for gaining practical experience. In the past, people have explored various places in the depths of the Amazon, the South Pole, the wilds of Africa, and so on, and vast knowledge can be acquired through such experiences. However, in some cases, such explorations may pose a risk to one's life. At such times, the use of telexistence is beneficial. Apparently, exploration by means of telexistence may also be used as a new form of leisure.

In the field of education, in addition to the utilization of a robot as a flight simulator for flight training, for practicing driving of a vehicle, etc., the robot may be used for an even higher level of training. In particular, at the time of driving or flying, it is highly effective to practice handling an aircraft or vehicle under extreme conditions or during accidents and breakdowns; however, the implementation of such practice is usually not possible, due to the inherent danger involved. Nevertheless, such training becomes possible by means of telexistence.

Further, if ordinary telexistence is coupled with computer-based virtual reality methods, various forms of real-time simulators could be constructed, and it would be possible to engineer the illusion of a person being within an almost-real artificial computer-generated environment.

This type of coupled telexistence has unlimited potential applications, one of them being gaming. In terms of educational applications, telexistence can be used to gain knowledge on physics and medicine, for example, by allowing a person to experience the world of elementary particles or to be one of the members of the microscopic exploration team depicted in the movie "Fantastic Voyage."

There is one more application of telexistence, resulting from a completely different perspective. In the future, society will probably leave dangerous, tedious, and unhygienic jobs to robots and machines. People will pursue more human and creative occupations. However, people dislike being made to behave in a particular manner. To begin with, who will assert that people should not be performing activities that are normally performed by robots? People must be able to do what they want as long as it does not disturb others (although this poses very difficult problems).

Therefore, what should be done when a person wants to perform an activity that is usually performed by robots? In practice, working in the same location as robots and machines is extremely dangerous. If accidents occur, it is not inconceivable that the functioning of society may be brought to a halt. This kind of activity may therefore exceed the bounds of liberty. Here again, telexistence provides a solution. In other words, it provides a method by which a person's desire to be involved in activities performed by robots and machines may be satisfied in a manner that does not infringe upon safety and social considerations.

8.5. Telexistence and Society

Let us consider the term "ubiquitous," which means to exist in every location or to appear to exist everywhere at the same time. A long time ago, during my high school student years, when I first encountered this word, it felt like scientifically unrealizable; at the same time, something about it touched me and it remained in my mind.

There is a demand for humans to be able to perform operations using their superior judgment and perspective in environments that are dangerous for them to work in directly, for example, environments with strong radiation, vacuum, high pressure, high temperature, and other severe conditions. If robots are distributed in dangerous environments, and whenever necessary during operations humans are able to take the place of these robots, perceive the environment from a safe remote environment, and control the robots with the sense that they themselves are acting directly, the advantages of machines may be combined with the superior judgment and perspective of humans in an ideal manner.

It was in the early autumn of 1980 that I conceived the concept of telexistence in order to actualize such actions and began research. At some point during the actual progress of the research, the long-forgotten word "ubiquitous" crossed my mind. This notion of telexistence can make humans virtually ubiquitous. That is, many intelligent robots are acting in all kinds of locations. If communication channels are established with some of these robots and they are operated using the method of telexistence control, the result is equivalent to the situation in which the person directly exists at the location of each robot.

By controlling robots sequentially or by controlling a number of them simultaneously, does it not appear that the person is ubiquitous? This is not merely a thought experiment confined to one's mind. In the midst of a series of experiences that I underwent during actual telexistence research, while controlling my own robot and being immersed in remote scenes with a sensation of presence, I remembered the term "ubiquitous" and was convinced that humans could become ubiquitous by using telexistence technology.

In a sense, robotics research is human research. The more one researches robotics, the more one becomes aware that there are many things to be learned from humans and living beings. The more one researches humans and living beings, the more one becomes overwhelmed by the precision and ingenuity of their mechanisms.

Conversely, when robots and mechanical functions are implemented, they become good models for understanding humans and living beings. That is to say, a tangible demonstration by means of such methods physically confirms the fact that certain human functions are achievable. Furthermore, since these physical demonstrations are created by people, they are precisely understood down to the fine details and can thus be used to construct convenient models that may be analyzed easily. In this sense, one may suppose that telexistence and other robot technologies will not only provide benefits to the society and lifestyles of humans in the future but will also have a significant influence on their culture and way of thinking.

Even so, these technologies must be for humans and must prioritize humans. By nature, the significance of robots lies in the fact that they act on behalf of humans. By compensating for human functions lost through misfortune, expanding the range of human capabilities through robot technology, and simultaneously benefiting from feedback that allows us to understand humans through robots, we may hope for greater developments in humans and the society.

We live in an era where the speed of technological advances has far exceeded the rate of individuals' adaptation and learning abilities. To avoid a fatal outcome, researchers and developers of computer technology must cultivate a human-centered paradigm. It is essential to establish a technology, using people who can interact naturally with machines such as robots and computers as if they were interacting with other people or nature, in a cybernetic way that is truly human-oriented, thoroughly backed up by knowledge of physiology and cognitive behavioral psychology. We can permit various forms of an interface; however, the default interface has to be object-oriented, with a clear standard by which people can interact with machines as if they were handling natural objects.

Telexistence, a virtual reality technology, runs concurrent with this new human-centered notion. A machine can improve not only the motor abilities but also the intellectual abilities of people in a very natural way. The focus of telexistence is always on the person, who feels as if he or she was voluntarily moving his or her body and using his or her brain, but in actuality, the machine is expanding his or her abilities and power.

To conclude, telexistence will liberate humans from the bonds of time and space in a human-centered manner.

Appendix A
Color Figures

Fig. 3.4.   Telesar (TELExistence Surrogate Anthropomorphic Robot).

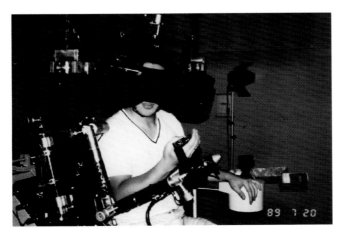

Fig. 3.5.   Telexistence master system.

Fig. 3.14.   Telexistence cockpit for controlling humanoid robots.

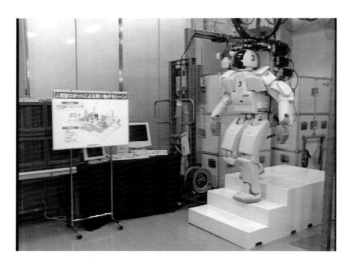

Fig. 3.15.   HRP humanoid robot at work.

Fig. 4.13. Prototype mobile telexistence vehicle (televehicle).

Fig. 4.14. Head-linked stereo display providing a sensation of presence.

Fig. 5.13.   Optical camouflage using RPT.

Fig. 5.15.   Transparent cockpit of a passenger car.

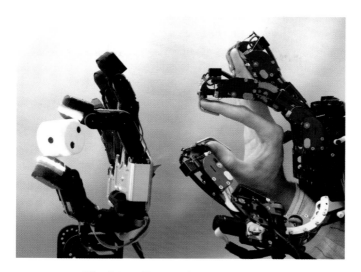

Fig. 6.25. Haptic telexistence system.

Fig. 6.27.   Examples of RPT appearance (left: TELESAR II without projection; center: with projection of an operator; right: operator at the controls).

Fig. 6.35.   Projected images of the operator captured from various angles.

Fig. 7.3.   360° panoramic image displayed in TWISTER II.

Fig. 7.4.   360° panoramic image displayed in TWISTER IV.

Fig. 7.13.   User observed from outside the TWISTER II booth.

Fig. 7.14.   User observed from outside the TWISTER IV booth.

Fig. 7.26.   Example of omnistereo images.

Bibliography

Akin, D. L., Minsky, M. L., Thiel, E. D. and Kurtzman, C. R. (1983). Space application of automation, robotics and machine intelligence systems (ARAM1S) — Phase II, Telepresence technology base development, NASA Contract Report, Number 3734, Washington, DC: National Aeronautics and Space Administration.

Anderson, R. J. and Spong, M. W. (1989). Bilateral control of teleoperation with time delay, *IEEE Transactions on Automatic Control*, Vol. AC-34, No. 5, pp. 494–501.

Arai, H., Tachi, S. and Miyajima, I. (1989). Development of a power-assisted head-coupled display system using a direct-drive motor, *Advanced Robotics*, Vol. 3, No. 2, pp. 123–130.

Brooks, Jr., F. P. (1986). Walkthrough-a dynamic graphics systems for simulating virtual buildings, *Proceedings of the ACM 1986 Workshop on Interactive 30 Graphics*, Chapel Hill, NC, October 1980, pp. 9–21.

Burdea, G., Zhuang, J., Roskos, E., Silver, D. and Langrana, N. (1992). A portable dextrous master with force feedback, *Presence*, Vol. 1, No. 1, pp. 18–27.

Charles, J. and Vertut, J. (1977). Cable controlled deep submergence teleoperator system, *Mechanism and Machine Theory*, pp. 481–492.

Cruz-Neira, C., Sandin, D. J. and DeFanti, T. A. (1993). Surrounded-screen projection-based virtual reality: The design and implementation of the CAVE, *Proceedings of the ACM SIGGRAPH'93*, pp. 135–142.

DeFanti, T. A., Leigh, J., Brown, M. D., Sandin, D. J., Yu, O., Zhang, C., Singh, R., He, E., Alimohideen, J., Krishnaprasad, N. K., Grossman, R., Mazzucco, M., Smarr, L., Ellisman, M., Papadopoulos, P., Chien, A. and Orcutt, J. (2003). Teleimmersion and visualization with the OptIPuter, in *Telecommunication, Teleimmersion and Telexistence*, Tachi, S. (ed.), IOS Press, ISBN 1-58603-338-7, pp. 25–71.

Fisher, S. S., Mcgreevy, M., Humphries, J. and Robinett, W. (1986). Virtual environment display system, *Proceedings of the ACM 1986 Workshop on Interactive 30 Graphics*, Chapel Hill, NC, October 1986, pp. 77–87.

Fjermedal, G. (1986). Part III: Outrageous world, Chapter 16: In the land of Zen and robots, in *The Tomorrow Makers*, Macmillan Publishing Company, ISBN 0-02-538560-7, pp. 232–235.

Fuchs, H. and Ackerman, J. (1999). Displays for augmented reality: Historical remarks and future prospects, in *Mixed Reality*, Tamura and Ohta (eds.), Springer-Verlag, pp. 31–40.

Halme, A. (2005). From teleoperation to cognition-based teleinteraction, in *Telecommunication, Teleimmersion and Telexistence II*, Tachi, S. (ed.), IOS Press, ISBN 1-58603-519-3, pp. 39–61.

Hayashi, J., Tanaka, K., Inami, M., Sekiguchi, D., Kawakami, N. and Tachi, S. (2002). Issues in image-capture system for TWISTER, *Proceedings of the ICAT2002, 12th International Conference on Artificial Reality and Tele-Existence*, Tokyo, Japan, pp. 44–51.

Hightower, J. D., Spain, E. H. and Bowles, R. W. (1987). Telepresence: A hybrid approach to high performance robots, *Proceedings of the International Conference on Advanced Robotics (ICAR '87)*, Versailles, France, pp. 563–573.

Hirose, M. (2003). Wearable computers and ubiquitous media space, in *Telecommunication, Teleimmersion and Telexistence*, Tachi, S. (ed.), IOS Press, ISBN 1-58603-338-7, pp. 133–163.

Hirose, M. (2005). Real-world virtual reality, in *Telecommunication, Teleimmersion and Telexistence II*, Tachi, S. (ed.), IOS Press, ISBN 1-58603-519-3, pp. 89–111.

Hirota, K. and Hirose, M. (1993). Development of surface display, *Proceedings of the 1993 IEEE Annual Virtual Reality International Symposium (VRAIS)*, Seattle, Washington, pp. 256–262.

Hoshino, H. and Tachi, S. (1998). A method to represent an arbitrary surface in encounter type shape representation system, *Proceedings of the 7th IEEE International Workshop on Robot and Human Communication (RO-MAN '98)*, Takamatsu, Japan, pp. 107–114.

Inami, M., Kawakami, N., Sekiguchi, D., Yanagida, Y., Maeda, T., Mabuchi, K. and Tachi, S. (1999). Head-mounted projector, *ACM SIGGRAPH'99 Conference Abstracts and Applications (Emerging Technologies)*, p. 179.

Inami, M., Kawakami, N., Sekiguchi, D., Yanagida, Y., Maeda, T. and Tachi, S. (2000). Visuo-haptic display using head-mounted projector, *Proceedings of the IEEE Virtual Reality 2000*, New Brunswick, New Jersey, USA, pp. 233–240.

Iwata, H. and Noma, H. (1990). Artificial reality with force feedback: Development of desktop virtual space with compact master manipulator, *ACM Computer Graphics*, Vol. 24, No. 4.

Kajimoto, H., Inami, M., Kawakami, N. and Tachi, S. (2004). SmartTouch: Electric skin to touch the untouchable, *IEEE Computer Graphics & Applications Magazine*, Vol. 24, No. 1, pp. 36–43.

Kamiyama, K., Kajimoto, H., Kawakami, N. and Tachi, S. (2004). Evaluation of a vision-based tactile sensor, *Proceedings of IEEE International Conference on Robotics and Automation*, Vol. 2, pp. 1542–1547.

Kato, N., Jo, K., Minamizawa, K., Kawakami, N. and Tachi, S. (2008). Omni-directional stereoscopic video camera system for visual telexistence, *Transactions of the Virtual Reality Society of Japan*, Vol. 13, No. 3, pp. 353–362 (in Japanese).

Kawabuchi, I., Tachi, S. and Kawakami, N. (2003). Rotation and extension/ retraction link mechanism, Japanese Patent filed on December 18, 2003.

Kawakami, N., Inami, M., Maeda, T. and Tachi, S. (1998). MediaX'tal — Projecting virtual environments on ubiquitous object-oriented retroreflective screens in the real environment, *SIGGRAPH'98*, Orlando, Florida, USA.

Kawakami, N., Sekiguchi, D., Kajimoto, H. and Tachi, S. (2005). Telesar-PHONE — Communication robot based on next generation telexistence technologies, *Proceedings of 36th International Symposium on Robotics (ISR2005)*, Tokyo, Japan.

Kuma, K. (2005). Digital gardening, in *Telecommunication, Teleimmersion and Telexistence II*, Tachi, S. (ed.), IOS Press, ISBN 1-58603-519-3, pp. 81–87.

Kunita, Y., Ogawa, N., Sakuma, A., Inami, M., Maeda, T. and Tachi, S. (2001). Immersive autostereoscopic display for mutual telexistence: TWISTER I (Telexistence Wide-angle Immersive STEReoscope Model I), *Proceedings of the IEEE Virtual Reality 2001*, Yokohama, Japan, pp. 31–36.

Luneburg, R. K. (1950). The metric of binocular visual space, *Journal of Optical Society of America*, Vol. 40, No. 10, pp. 627–642.

Maeda, T. and Tachi, S. (1992). Development of light-weight binocular head-mounted displays, *Proceedings of the Second International Symposium on Measurement and Control in Robotics (ISMCR '92)*, Tsukuba, Japan, November 1992, pp. 281–288.

Mann, R. W. (1965). The evaluation and simulation of mobility aids for the blind, *American Foundation Blind Research Bulletin*, No. 11, pp. 93–98.

McNeely, W. A. (1993). Robotic graphics: A new approach to force feedback for virtual reality, *Proceedings of the 1993 IEEE Virtual Reality Annual International Symposium (VRAIS)*, Seattle, Washington, pp. 336–341.

McRuer, D. T. and Jex, H. R. (1967). A review of quasi-linear pilot models, *IEEE Transactions on Human Factors Engineering*, Vol. HFE-8, pp. 241–249.

Minamizawa, K., Kajimoto, H., Kawakami, N. and Tachi, S. (2007). Wearable haptic display to present gravity sensation — Preliminary observations and device design, *Proceedings of the 2nd Joint Euro-haptics Conference and Symposium on Haptic Interfaces for Virtual Environment and Teleoperator Systems (World Haptics 2007)*, Tsukuba, Japan, pp. 133–138.

Minsky, M. (1980). Telepresence, *Omni*, Vol. 2, No. 9, pp. 44–52.

Naemura, T. and Harashima, H. (2003). Ray-based approach to integrated 3D visual communication, in *Telecommunication, Teleimmersion and Telexistence*, Tachi, S. (ed.), IOS Press, ISBN 1-58603-338-7, pp. 73–97.

Naemura, T. (2005). Human communication media, in *Telecommunication, Teleimmersion and Telexistence II*, Tachi, S. (ed.), IOS Press, ISBN 1-58603-519-3, pp. 153–164.

Nakagawara, S., Tadakuma, R., Kajimoto, H., Kawakami, N. and Tachi, S. (2004). A method to solve inverse kinematics of redundant slave arm in the master–slave system with different degrees of freedom, *Proceedings of the 2004 International Symposium on Measurement, Control, and Robots*, NASA Johnson Space Center, Houston, Texas, USA.

Nakagawara, S., Kawabuchi, I., Kajimoto, H., Kawakami, N. and Tachi, S. (2005). An encounter-type multi-fingered master hand using circuitous joints, *Proceedings of the IEEE International Conference on Robotics and Automation (ICRA2005)*, Barcelona, Spain, pp. 2667–2672.

Nishiyama, N., Hoshino, H., Sawada, K., Baba, A., Sekine, T., Yamada, W., Terasawa, A., Nakajima, R., Tokunaga, Y. and Yoneda, M. (2003). Communication agent embedded in humanoid robot, *Proceedings of the SICE Annual Conference 2003 (SICE2003)*, pp. 1342–1347.

Oyama, K., Tsunemoto, N., Tachi, S. and Inoue, T. (1993). Experimental study on remote manipulation using virtual reality, *Presence*, Vol. 2, No. 2, pp. 112–124.

Pepper, R. L., Cole, R. E. and Spain, E. H. (1983). The influence of camera and head movement on perceptual performance under direct and TV-displayed conditions, *Proceedings of the SID*, Vol. 24, No. 1, pp. 73–80.

Rheingold, H. (1991). Chapter 12: Out-of-the-body: Tuning in to the tele-existence, in *Virtual Reality, Touchstone*, ISBN 0-671-77897-8, pp. 258–269.

Satava, R. M. (2005). Telemedicine: Virtual reality, and other technologies that will transform how healthcare is provided, in *Telecommunication, Teleimmersion and Telexistence II*, Tachi, S. (ed.), IOS Press, ISBN 1-58603-519-3, pp. 63–79.

Sato, K., Minamizawa, K., Kawakami, N. and Tachi, S. (2007). Haptic telexistence, *34th International Conference on Computer Graphics and Interactive Techniques (ACM SIGGRAPH 2007)*, San Diego, USA.

Schmandt, C. (1983). Spatial input/display correspondence in a stereoscopic computer graphic work station, *Computer Graphics*, Vol. 17, No. 3, pp. 253–261.

Sheridan, T. B. and Ferrell, W. R. (1974). *Man–Machine Systems*. Cambridge, MA: MIT Press.

Shimamoto, M. S. (1992). Teleoperator/telepresence system (TOPS) concept verification model (CVM), *Development in Recent Advances in Marine Science and Technology '92*, Saxena, N. K. (ed.), pp. 97–104.

Sonoda, T., Endo, T., Suzuki, Y., Kawakami, N. and Tachi, S. (2005). X'tal Visor, *ACM SIGGRAPH 2005 (Emerging Technologies)*.

Stark, L., Kim, W. S., Tendick, F., Hannaford, B., Ellis, S. *et al.* (1987). Telerobotics: Display, control and communication problems, *IEEE Journal of Robotics and Automation*, Vol. RA-3, No. 1, pp. 67–75.

Sutherland, I. E. (1968). A head-mounted three dimensional display, *Proceedings of the Fall Joint Computer Conference*, pp. 757–764.

Tachi, S., Tanie, K. and Komoriya, K. (1980). Evaluation apparatus of mobility aids for the blind, Japanese Patent 1462696, 26 December 1980.

Tachi, S., Tanie, K. and Komoriya, K. (1981a). An operation method of manipulators with functions of sensory information display, Japanese Patent 1458263, 11 January 1981.

Tachi, S., Tanie, K., Komoriya, K., Hosoda, Y. and Abe, M. (1981b). Guide dog robot: Its basic plan and some experiments with MELDOG MARK I, *Mechanism and Machine Theory*, Vol. 16, No. 1, pp. 21–29.

Tachi, S. and Abe, M. (1982). Study on tele-existence (I): Design of visual display, *Proceedings of the 21st Annual Conference of the Society of Instrument and Control Engineers (SICE)*, Tokyo, Japan, pp. 167–168 (in Japanese).

Tachi, S. and Komoriya, K. (1982). The third generation robotics, *Journal of SICE*, Vol. 21, No. 12, pp. 1140–1146 (in Japanese).

Tachi, S., Mann, R. W. and Rowell, D. (1983). Quantitative comparison of alternative sensory displays for mobility aids for the blind, *IEEE Transactions on Biomedical Engineering*, Vol. BM-30, No. 9, pp. 571–577.

Tachi, S., Tanie, K., Komoriya, K. and Kaneko, M. (1984). Tele-existence (I): Design and evaluation of a visual display with sensation of presence, *Proceedings of the 5th Symposium on Theory and Practice of Robots and Manipulators (RoManSy '84)*, Udine, Italy, London: Kogan Page, pp. 245–254.

Tachi, S. and Komoriya, K. (1985). Guide dog robot, in Brady, M. *et al.* (eds.), *The Robotics Research 2, The Second International Symposium*, MIT Press, pp. 333–349.

Tachi, S., Tanie, K., Komoriya, K. and Abe, M. (1985). Electrocutaneous communication in a guide dog robot (MELDOG), *IEEE Transactions on Biomedical Engineering*, Vol. BME-32, No. 7, pp. 461–469.

Tachi, S., Arai, H. and Maeda, T. (1988a). Tele-existence simulator with artificial reality. (1): Design and evaluation of a binocular visual display using solid models, *Proceedings of the IEEE International Workshop on Intelligent Robots and System (IROS '88)*, Tokyo, Japan, October 1988, pp. 719–724.

Tachi, S., Arai, H., Morimoto, I. and Seet, G. (1988b). Feasibility experiments on a mobile tele-existence system, *Proceedings of the International Symposium and Exposition on Robots (19th ISIR)*, Sydney, Australia, November 1988, pp. 625–636.

Tachi, S., Arai, H. and Maeda, T. (1989). Development of an anthropomorphic tele-existence slave robot, *Proceedings of the International Conference on Advanced Mechatronics (ICAM)*, Tokyo, Japan, May 1989, pp. 385–390.

Tachi, S., Arai, H. and Maeda, T. (1990). Tele-existence master–slave system for remote manipulation, *Proceedings of the IEEE International Workshop on Intelligent Robots and Systems (IROS '90)*, Tsuchiura, Japan, pp. 343–348.

Tachi, S., Arai, H., Maeda, T., Oyama, E., Tsunemoto, N. and Inoue, Y. (1991a). Tele-existence experimental system for remote operation with a sensation of presence, *Proceedings of the International Symposium on Advanced Robot Technology ('91 ISART)*, Tokyo, Japan, March, pp. 451–458.

Tachi, S., Sakaki, T., Arai, H., Nishizawa, S. and Pelaez-Polo, J. F. (1991b). Impedance control of a direct-drive advanced manipulator without using force sensors, *Advanced Robotics*, Vol. 5, No. 2, pp. 183–205.

Tachi, S. and Sakaki, T. (1992). Impedance controlled master–slave manipulation system: Part I. Basic concept and application to the system with a time delay, *Advanced Robotics*, Vol. 6, No. 4, pp. 483–503.

Tachi, S. and Yasuda, K. (1994). Evaluation experiments of a tele-existence manipulation system, *Presence*, Vol. 3, No. 1, pp. 35–44.

Tachi, S., Maeda, T., Hirata, R. and Hoshino, H. (1994). A construction method of virtual haptic space, *Proceedings of the 4th International Conference on Artificial Reality and Tele-Existence (ICAT '94)*, Tokyo, Japan, July 1994, pp. 131–138.

Tachi, S., Maeda, T., Yanagida, Y., Koyanagi, M. and Yokoyama, H. (1996). A method of mutual tele-existence in a virtual environment, *Proceedings of the 6th International Conference on Artificial Reality and Tele-Existence, (ICAT '96)*, Makuhari, Chiba, Japan, pp. 9–18.

Tachi, S. (1998). Real-time remote robotics toward networked telexistence, *IEEE Computer Graphics and Applications*, Vol. 18, pp. 6–9.

Tachi, S. (1999a). Toward next generation telexistence, *Proceedings of IMEKO-XV World Congress*, Vol. X (TC-17 & ISMCR '99), Tokyo, Osaka, Japan, pp. 173–178.

Tachi, S. (1999b). Augmented telexistence, in *Mixed Reality*, Tamura and Ohta (eds.), Springer-Verlag, pp. 251–260.

Tachi, S. (2001). Toward the telexistence next generation, *Proceedings of the 11th International Conference on Artificial Reality and Tele-Existence (ICAT2001)*, Tokyo, Japan, pp. 1–8.

Tachi, S., Komoriya, K., Sawada, K., Nishiyama, T., Itoko, T., Kobayashi, M. and Inoue, K. (2001). Development of telexistence cockpit for humanoid robot control, *Proceedings of the 32nd International Symposium on Robotics (ISR2001)*, Seoul, Korea, pp. 1483–1488.

Tachi, S. (2003a). Two ways of mutual telexistence: TELESAR and TWISTER, in *Telecommunication, Teleimmersion and Telexistence*, Tachi, S. (ed.), IOS Press, ISBN 1-58603-338-7, pp. 3–24.

Tachi, S. (2003b). Telexistence and retroreflective projection technology (RPT), *Proceedings of the 5th Virtual Reality International Conference (IVRIC2003)*, Laval Virtual, France, pp. 69/1–69/9.

Tachi, S., Komoriya, K., Sawada, K., Nishiyama, T., Itoko, T., Kobayashi, M. and Inoue, K. (2003). Telexistence cockpit for humanoid robot control, *Advanced Robotics*, Vol. 17, No. 3, pp. 199–217.

Tachi, S., Kawakami, N., Inami M. and Zaitsu, Y. (2004). Mutual telexistence system using retroreflective projection technology, *International Journal of Humanoid Robotics*, Vol. 1, No. 1, pp. 45–64.

Tachi, S. (2005). Telexistence: Next-generation networked robotics, in *Telecommunication, Teleimmersion and Telexistence II*, Tachi, S. (ed.), IOS Press, ISBN 1-58603-519-3, pp. 3–38.

Tachi, S. (2007). TWISTER: Immersive omnidirectional autostereoscopic 3D booth for mutual telexistence, *Proceedings of ASIAGRAPH 2007*, Vol. 1, No. 2, Tokyo, Japan, pp. 1–6.

Tachi, S., Kawakami, N., Nii, H., Watanabe, K. and Minamizawa, K. (2008). TELEsarPHONE: Mutual telexistence master–slave communication system based on retroreflective projection technology, *SICE Journal of Control, Measurement, and System Integration*, Vol. 1, No. 5, pp. 1–10.

Tadakuma, R., Asahara, Y., Kajimoto, H., Kawakami, N. and Tachi, S. (2005). Development of anthropomorphic multi-D.O.F. master–slave arm for mutual telexistence, *IEEE Transactions on Visualization and Computer Graphics*, Vol. 11, No. 6, pp. 626–636.

Tanaka, K., Hayashi, J., Inami, M. and Tachi, S. (2001). The design and development of TWISTER II: Immersive full-color autostereoscopic display, *Proceedings of the ICAT2001, 11th International Conference on Artificial Reality and Telexistence*, Tokyo, Japan, pp. 56–63.

Tanaka, K., Hayashi, J., Inami, M., Kunita, Y., Maeda, T. and Tachi, S. (2002). TWISTER: A media booth, *Emerging Technologies, SIGGRAPH 2002*, San Antonio, USA.

Tanaka, K., Hayashi, J., Inami, M. and Tachi, S. (2004). TWISTER: An immersive autostereoscopic display, *Proceedings of the IEEE Virtual Reality 2004*, Chicago, USA.

Tanaka, K. and Tachi, S. (2005). TORNADO: Omnistereo video imaging with rotating optics, *IEEE Transactions on Visualization and Computer Graphics*, Vol. 11, No. 6, pp. 614–625.

Turing, A. M. (1950). Computing machinery and intelligence, *Mind*, Vol. 59, pp. 433–460.

Watanabe, K., Kawabuchi, I., Kawakami, N., Maeda, T. and Tachi, S. (2008). TORSO: Development of a telexistence visual system using a 6-D.O.F. robot head, *Advanced Robotics*, Vol. 22, pp. 1053–1073.

Yanagida, Y. and Tachi, S. (1993). Virtual reality system with coherent kinesthetic and visual sensation of presence, *Proceedings of the 1993 JSME International Conference on Advanced Mechatronics (ICAM)*, Tokyo, Japan, pp. 98–103.

Yokokohji, Y., Hollis, R. L. and Kanade, T. (1996). What you can see is what you can feel, *Proceedings of the IEEE Virtual Reality Annual International Symposium (VRAIS '96)*, Santa Clara, CA, USA, pp. 46–53.

Yoshida, T., Jo, K., Minamizawa, K., Nii, H., Kawakami, N. and Tachi, S. (2008). Transparent cockpit: Visual assistance system for vehicle using retroreflective projection technology, *Proceedings of the IEEE Virtual Reality 2008*, Reno, USA, pp. 185–188.

van Dam, A., Fuchs, H., Becker, S., Holden, L., Ilie, A., Low, K. L., Spalter, A. M., Yang, R. and Welch, G. (2003). Immersive electronic books for teaching surgical procedures, in *Telecommunication, Teleimmersion and Telexistence*, Tachi, S. (ed.), IOS Press, ISBN 1-58603-338-7, pp. 99–132.

Welch, G., Yang, R., Cairns, B., Towles, H., State, A., Ilie, A., Becker, S., Russo, D., Funaro, J., Sonnenwald, D., Mayer-Patel, K., Allen, B. D., Yang, H., Freid, E., van Dam, A. and Fuchs, H. (2005). 3D telepresence for off-line surgical training and on-line remote consultation, in *Telecommunication, Teleimmersion and Telexistence II*, Tachi, S. (ed.), IOS Press, ISBN 1-58603-519-3, pp. 113–152.

About the Author

 Susumu Tachi received his B.E., M.S. and Ph.D. degrees from The University of Tokyo, in 1968, 1970 and 1973, respectively.

He joined the Faculty of Engineering of The University of Tokyo in 1973, and in 1975, he moved to the Mechanical Engineering Laboratory, Ministry of International Trade and Industry, Japan, where he served as the Director of the Biorobotics Division. From 1979 to 1980, Dr. Tachi was a Japanese Government Award Senior Visiting Scientist at the Massachusetts Institute of Technology, USA. In 1989, he rejoined The University of Tokyo, and served as Professor at the Department of Information Physics and Computing till March 2009. He is currently Professor Emeritus of The University of Tokyo and Professor of Robotics and Virtual Reality at Keio University, Japan.

He is the 46th President of the Society of Instrument and Control Engineers (SICE), a Founding Director of the Robotic Society of Japan (RSJ), and the Founding President of the Virtual Reality Society of Japan (VRSJ). From 1988, he has served as Chairman of the IMEKO Technical Committee on Measurement in Robotics and directed the organization of ISMCR symposia. He initiated and founded the International Conference on Artificial Reality and Telexistence (ICAT) and International-collegiate Virtual Reality Contest (IVRC). He is a member of IEEE VR Steering Committee, and served as General Chair of IEEE Virtual Reality Conferences.

Index